WOMEN AND GIS

Mapping Their Stories

Written by Esri Press

Esri Press
REDLANDS | CALIFORNIA

Compass roses throughout: CloudyStock, Makhnach_S, Vector Tradition/Shutterstock
Map line art: Al-xVadinska/Shutterstock

Esri Press, 380 New York Street, Redlands, California 92373-8100

23 22 21 20 19   1 2 3 4 5 6 7 8 9 10

Printed in the United States of America

Library of Congress Cataloging-in-Publication Data

Names: Esri Press, issuing body.
Title: Women and GIS : mapping their stories.
Description: Redlands, Calif. : Esri Press, 2019.
Identifiers: LCCN 2018036037 (print) | LCCN 2018038546 (ebook) | ISBN 9781589485297 (electronic) | ISBN 9781589485280 (hardcover : alk. paper)
Subjects: LCSH: Women geographers--Biography. | Geographic information systems.
Classification: LCC G67 (ebook) | LCC G67 .W66 2019 (print) | DDC 910.92/52--dc23
LC record available at https://urldefense.proofpoint.com/v2/url?u=https-3A__lccn.loc.gov_2018036037&d=DwIFAg&c=n6-cguzQvX_tUIrZOS_4Og&r=qNU49__SCQN30XC-f38qj8bYYMTIH4VCOt2Jb8fvjUA&m=F8UqfvbPsLE8YcR5fl7sC_F08ciLt4tr2I6jVCbFPD8&s=lGsShtIj9WvLf4mqxS4WmrQeGknp-5FKafr0c25iyYQ&e=

Ask for Esri Press titles at your local bookstore or order by calling 1-800-447-9778. You can also shop online at www.esri.com/esripress. Outside the United States, contact your local Esri distributor or shop online at eurospanbookstore.com/esri.

Esri Press titles are distributed to the trade by the following:

In North America:
Ingram Publisher Services
Toll-free telephone: 800-648-3104
Toll-free fax: 800-838-1149
E-mail: customerservice@ingrampublisherservices.com

In the United Kingdom, Europe, the Middle East and Africa, Asia, and Australia:
Eurospan Group
3 Henrietta Street
London WC2E 8LU
United Kingdom
Telephone 44(0) 1767 604972
Fax: 44(0) 1767 6016-40
E-mail:eurospan@turpin-distribution.com

Dedication

We at Esri Press dedicate this book to all the women right here at Esri who bring their special knowledge, intelligence, and life experience to help craft cutting-edge technology each day and to render their support in advancing GIS and The Science of Where®.

Because we couldn't possibly choose between all these many doers and achievers, we wish to express our appreciation and thanks for all these women, for their talents and for their hard work and integrity in making this a better Esri. And we encourage all the other women out there yearning to make their way to take inspiration from the women featured in this book and to "strive for excellence and never give up your dreams," to "be honest and thoughtful, and then make a difference," to be "courageous and creative," to "never stop being in wonder of your work," and above all, to "not be afraid to fail."

We urge the women who are still rising to the fore to take the dare and be open to what life may bring—both the challenges and the rewards. And thus, to celebrate the triumph in doing the type of work it is that you love and enjoy.

Contents

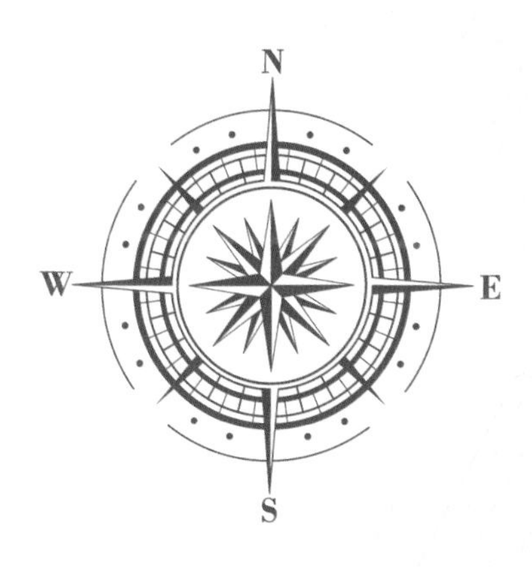
N
W
E
S

Foreword

The release of this book could not come at a better time, a time that ranks among the most significant in terms of the sheer power and impact of women's voices.

The women's movement has sparked a cultural conversation, encouraging people to face their own beliefs and behavior, especially in the workplace.

More women are running for elected office than ever before, and from a variety of professions, from nursing, to geology, to naval aviation.

Lehua Kamalu became the first woman to serve as both captain and lead navigator of a Polynesian voyaging canoe, successfully guiding the *Hikianalia* over 2,800 nautical miles from Honolulu, Hawaii, to Half Moon Bay, California, in 23 days.

And perhaps to top it off, two women received science Nobel Prizes in 2018: Donna Strickland in physics—only the third female winner ever, and the first woman to win it in 55 years—and Frances Arnold in chemistry—only the fifth female winner after Ada Yonath in 2009.

As Donna noted upon receiving her prize, "We need to celebrate women physicists because they're out there. ... I'm honored to be one of those women."

So, too, is the case with GIS. We need to celebrate the women of GIS because they're also out there. And they're exceedingly proud and honored to be part of a field and a technology that is literally saving the world. And more women are coming into the field all the time, bringing with them powerful and positive voices, diverse perspectives, and surgical insights. They're exerting influence within the GIS industry, as well as in the worlds of academia, STEM education, government, the humanitarian nonprofit space, exploration, conservation, even the Catholic Church.

This book is a wonderful sampling of these prescient forces. These are the stories of innovators, leaders, explorers, teachers, mentors, doers (!), many of whom have been hidden figures for far too long. But no more. I hope that you are as blown away by these profiles as I am. The quotations from these women alone are an absolute treasure trove. And while these are stories for GIS veterans and newcomers alike, both women and men, may this book be especially inspiring for female newcomers to the world of GIS.

—Dawn J. Wright,
Esri Chief Scientist

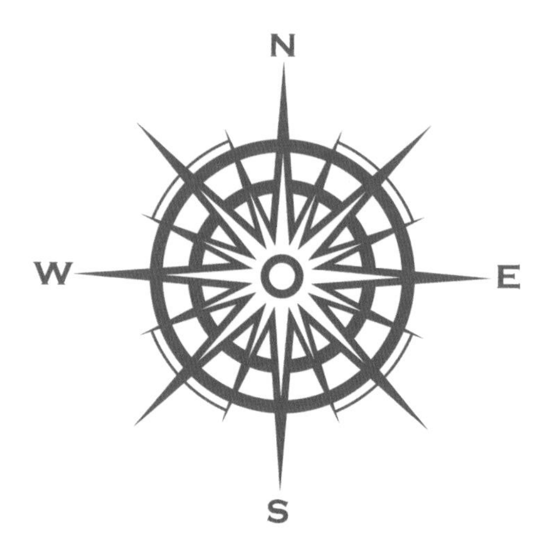

Preface

In early 2018, Stacy Krieg (my acquisitions manager) and I were given the extraordinary opportunity of spending an hour with Kathryn D. Sullivan (one of the women profiled in this book). We wanted her thoughts about how we could be part of bringing the knowledge and use of GIS to everyone. How we could reach beyond our existing users to people of any age, walk of life, job, or interest. How we could show them that applying data science to any problem or idea could improve the outcome. Things from better predictive models for wildfires to save lives, to real-time evacuation routes for hurricanes, to finding the best location for a sports park.

As we talked, Kathryn guided us to the realization that perhaps the most important group we should focus on is young people. They are the future generation(s) that will take on the challenges of the world. She said, speak to them about what interests them, not about the technology. Most are passionate about making the world better and improving the lives of those around them. Show them how mapping and GIS can help their passions become reality. Encourage them to take science, technology, engineering, and math (STEM) classes. Break the misconception that STEM is hard and not for girls. Break the "I can't do that" mindset.

When I left the meeting, I continued to think about what we'd discussed. It made me think of my own daughter. She has always had a natural talent for engineering and math. When she was 8 and 9 years old, you could place a box of gears, switches, and other parts in front of her, and she'd build you things. Complex things. No directions, no guidance. She'd say she just saw it in her mind and knew how to build it. She's still that way today, in high school. So you can imagine my shock every new school year when she comes home stressed about how difficult the new biology, math, physics—whatever—class is going to be, just because that's what her peers are saying about it. Even a young girl who was building adult-level construction sets by age 11 is at risk of falling victim to the myth that STEM is hard, especially for girls.

My daughter is lucky enough to have people around her who help her break her fears and have confidence in her ability to do whatever she sets her mind to, but not all young people have that support. Someone who will tell them they can when they think they can't. Someone who will tell them it doesn't matter where you come from or the struggles you face. Someone who will tell them it's where you're going that matters, and that you should believe in yourself.

Having that type of support then made me think of my own mother. She was a stay-at-home mom, and she was (and still is) exceptional at that job. She was the person who taught me that I could do anything, be anything, and that I was strong enough to face anything. She insisted I take every science and math class I could squeeze into a schedule. The term STEM wasn't used back then, but she knew the value in it, and she trained me to never be afraid of it. She was the one who encouraged me to go back to school for GIS. She knew nothing about GIS but had read a magazine article about it. She felt so strongly that it was my future that she and my grandmother brought me the magazine. I didn't always listen to these two women, but on that occasion, I did. Now, 25-plus years later, I can say that I still thank them both, every day.

It was the influence of these women in my life, and my desire to be the same influence in my daughter's life, that gave me the idea for this book. I wanted to show that it didn't matter who you were, where you came from, or the struggles you faced. I wanted people to read about women from different backgrounds, different countries, and different generations who faced struggles along the way but never gave up on their dreams. Women who use science, technology, engineering, and math to make the world, and the lives of the people around them, better. If these women can do it, so can they.

I took my book idea to Stacy. I trust her opinion and honesty and her extensive experience in the publishing world. I also know she wants to have the same influence on her youngest daughter (also in high school) as I want to have on mine. She saw the vision and the goal of the book—to show young people someone like themselves, and then they might begin to believe that they can do it, too. When they believe they can, the ability to learn the technology and methodology comes easier. And so began our journey together creating (eventually, along with all the other amazing people we work with at Esri Press) a book that maps the life stories of the women in it, and hopefully inspires others to begin mapping their life stories, too.

—Catherine Ortiz
Manager and publisher, Esri Press

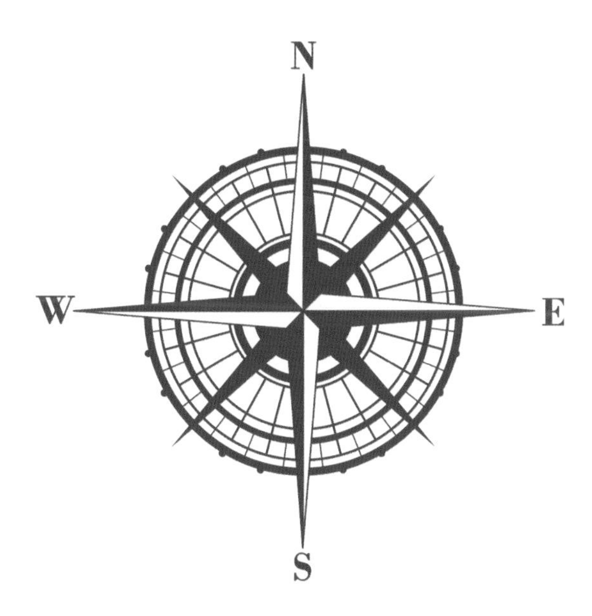

Women and GIS

Mapping Their Stories

Catherine Ball

Rising high with 'drones for good'

Catherine has won many awards for her work in business and education. The awards have opened networks, expanding her connections to mentors, sponsors, and business partners.

ON ONE OF HER FIRST PROJECTS AS AN ENVIRONMENTAL CONSULTANT on Australia's West Coast, Dr. Catherine "Cathy" Ball was struck by how difficult it was for scientists to monitor various animals on land and in the sea. It occurred to her that scientists should be able to get their data more easily and more safely. Perhaps she was inspired by the aerial images from David Attenborough's documentaries of the natural world that she had enjoyed as a child. Perhaps it seems like an obvious use now but using drones for this research hadn't been done before.

In 2013, Cathy envisioned reconnaissance aircraft that was being used to search for insurgents in Afghanistan instead being used to track and monitor turtle colonies. She educated stakeholders and overcame naysayers so that her team could fly human-size remotely piloted aircraft systems (RPAS), commonly called *drones*, hundreds of kilometers to track turtle habitats off the west Australian coast and, in the process, spotted endangered animals such as oceanic manta rays not seen by researchers in years. She learned a lot about what kinds of people she needed on her team to complete projects such as this one successfully. Now she leads five start-ups in Australia, all driven by values, with plans to go global.

She has come a long way from Nuneaton, a small town in the West Midlands region of England. Raised in a single-parent household, Cathy credits her mother with inspiring in her a strong sense of independence and hard work. At 21, her mother was hired by British Leyland Motor Corporation, the state-run manufacturer of Jaguar and Land Rover. In the 1970s, her mother received an offer letter stating her pay and conditions. Salary amounts were in a table separated into columns: one showing the rate for men and one for women. At a young age, Cathy was indignant about the gender pay discrepancy. But even her early exposure to pay inequities didn't deter her motivation to do her best, and today she says she makes sure that discrimination against women doesn't occur in any of her companies.

Cathy always excelled in the sciences, studying physics, biology, and chemistry, as well as French, and getting the highest grade in the class in geography, but she was disappointed when she had to give up the arts. The educational system did not allow her to pursue both, and she sees that restriction as a weakness in the system. Today when she holds meetings with stakeholders, she always wants to have an artist at the table.

Leonardo da Vinci demonstrates to us that science is creative, and scientists always do better work when they approach their problems creatively.

As she was growing up, certain events affected Cathy and influenced her life choices. Watching *Live Aid* on television as a girl made Cathy wonder how there could be such famine and suffering in the modern world. She thought she would go into medicine but changed her mind later during her gap year between high school and college. In the late 1990s in Zambia, Cathy witnessed what seemed like an entire generation disappear because of the spread of the killer disease AIDS. She realized that medicine wasn't enough and decided that environmental health was a better option for people and the planet. She was also influenced by the Piper Alpha oil rig explosion, which killed 167 workers in July 1988 in the North Sea off the coast of Scotland. Primarily caused by human error and miscommunication, it remains the world's deadliest oil rig accident and was the cause of improved safety regulations. Her father had been working on another oil rig in the North Sea at the time, and the disaster underscored Cathy's commitment to the environment and to working safely.

Cathy attended the University of Newcastle-upon-Tyne (now Newcastle University) in the United Kingdom (UK), obtaining a bachelor of science with honors in environmental protection and a PhD in spatial ecology, and descriptive and predictive statistics. Her mother worked hard to send her to school, where Cathy always held two part-time jobs to cover fees and living expenses. Once Cathy completed her education, she had to find a job to pay off her debt, and it looked as if becoming an environmental consultant for a corporation was the way to go. She says, "At that point, debt was making the choices for me."

During the next 10 years, Cathy worked as an environmental consultant with various corporations in the UK and Australia. Her skills and qualifications were equivalent to a passport, and, as a single woman, she thought "now or never" and made the leap across the world to Australia. Her innovation to use drone technology led to a promotion that allowed her to move across Australia to Queensland, a region that is now a global leader in the use of drone technology, spurred by Australia's airspace regulations that allow for developing smarter drone operations commercially.

[US President] John F. Kennedy once said, "Conformity is the jailer of freedom and the enemy of growth." I was taught to conform, but I've always been a free spirit at heart. [At that job] I had mentally left the building, and I no longer wanted to be in the system. I realized the only thing I could rely on was myself. I had been using my brain and my brand. But I wanted to be more than a replica of the system that I left. So I took my brain, my brand, my heart, and my gut, plus my ideas about how the world could be a better place, and I found myself on a winding, bumpy path of running five start-ups.

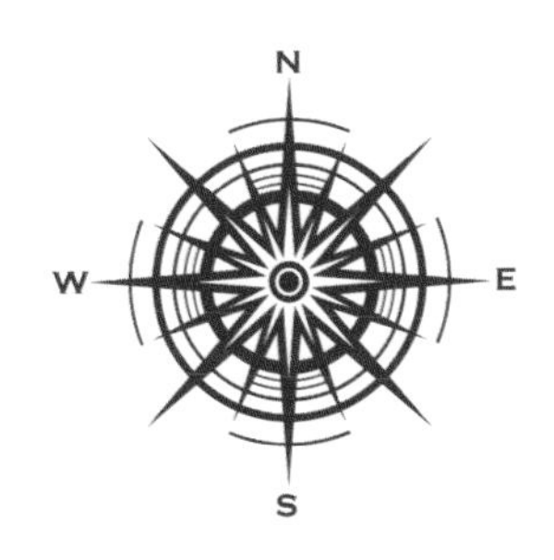

Although successful in various consulting positions, Cathy felt confined. She was still paying off her student debt, and although she wasn't climbing the corporate ladder and wasn't challenged by the work, she still thought it was the smartest way to go forward. Shortly after being awarded 2015 Telstra Queensland Business Woman of the Year for her innovation with drones, Cathy was laid off from her job. She recalls one sleepless night, getting up and staring in the mirror, and at that moment, she suddenly understood the paradox of an immovable object meeting an unstoppable force and saw clearly that she was letting fear make her decisions for her. She knew then and there that she could rely only on herself. Going back to bed, she was able to get some sleep—and hasn't looked back since.

Entrepreneurs need to be fleet, active, prescient, and willing to give it a go. Risk is just part of it. I am more fearless now than ever before.

Cathy is a scientist who became an entrepreneur, speaker, author, and innovator. Being an entrepreneur gives her more choices, she says, and she likes having control over her life.

Cathy currently manages five start-up businesses: Remote Research Ranges (RRR), the World of Drones Congress, World of Drones Education, the *Gumption Trigger* book project, and One Planet Woman. All are values-based businesses, which means they uphold integrity for the planet, Cathy says.

Cathy founded and is executive director of RRR, an advisory firm to international businesses, state, federal, and local governments, and schools, charities, universities, and parks. The firm advises on drone technology, big data management, and geoethics, the ethics around geospatial data. An expert in understanding drone technology, Cathy says the important thing is how the technology is used. She stands firmly for the #dronesforgood movement, whether it's bringing food to devastated villages in war-torn or disaster-affected countries, monitoring power lines, seeking people lost after a cyclone, or rescuing swimmers at sea.

Image by AustralianMiniadventures

Cathy leads the World of Drones Congress, which hosted over 1,200 delegates in 2018. The Premier of Queensland launched Australia's first statewide drone strategy. Additionally, the Australian National Drone Safety Forum was initiated.

The World of Drones Congress, first held in Brisbane, Australia, in 2017, supports the growing drone economy across the Asia Pacific region. Frustrated by the lack of business conversations around drone technology, Cathy envisioned a business, legal, investment, and ethics conference to help professionalize the drone industry by focusing on how drone technology could be used to keep people safe and generate new sources of revenue. She met with business partners who agreed with her vision and thus started the conference to connect people and facilitate the industry's development. Cathy expects the congress to be long running and expand internationally. World of Drones Education is the education and training start-up app related to the congress.

Image by AustralianMiniadventures

Educating girls and boys about drones is important to Cathy. She wants to see a world where just as many females as males are engaged in science, technology, engineering, and math (STEM). Currently women make up less than 1 percent of the global drone industry.

In 2016, Cathy crowdfunded and curated *Gumption Trigger*, a collection of stories by Australian women who have overcome obstacles through determination and gumption. Their stories of resilience serve to inspire other women working to realize their own dreams.

Through One Planet Woman, Cathy advises other start-up businesses and arranges speaking engagements.

Cathy wants to change the world, even though people tell her she can't.

If you believe you can't change the world, you certainly won't. I feel a responsibility to make the world a better place. I have heard that people have more regrets in life for things they haven't done, so I am doing all I can, even if I fail or make mistakes. My motto is: She who dares, learns. We can create change from outside the system that influences the system.

In 2017, revenue from her start-ups exceeded $1 million.

Sometimes Cathy finds she needs to be patient while those around her catch up to her vision. But it is worth the effort, she says, because she knows that she will not go far without the collaboration of others.

She and her husband are busy raising their infant son—it could almost get overwhelming, she says, but Cathy has a great reservoir of energy and vision that keeps her going. She imagines that her 12-year-old self would have been astonished to see her now. Young Cathy would not have allowed herself to dream quite this big. As an adult—as an accomplished speaker and networker, an innovator, a creator, a champion of social justice and the environment—Cathy is creating opportunities to change the world. She wants to leave some good behind and feels confident she will.

She reminds us:

All of us need to support the outliers because they are the ones that create a better future. ✱

Ranu Basu

Seeking human interconnectedness on a global scale

IN HER WORK, RANU BASU STRIVES TO BRING LIGHT TO COMPLEX QUESTIONS related to contested spaces of power and resistance. She unravels the geographies of communities that are displaced or marginalized because of systemic inequities and disparate levels of power. As an urban, social, and political geographer, she has lived and engaged with a global community and is deeply committed to understanding the stories, maps, and data that are part of broader social and transformational change. As a feminist geographer, she understands and communicates the values of diversity, power, and activism in her work and lives her life immersed in and inspired by the experiences of many different communities.

Ranu has long been fascinated by the mysteries of geography and the power of maps as discursive and strategic instruments for change.

Because of the nature of her father's profession working in a multinational company while she was a child, Ranu and her family moved frequently—from Canada to Peru to different parts of India and then back to Canada. Ranu never attended a school for more than three years at a time, and it seemed as if she was always going through a difficult process of adjustment and readjustment while she learned to read and write in a new language, make new friends, and adapt to new cultures. Yet this dislocation helped her learn

Whether it relates to the complexities of physical landscapes, environmental challenges to climate change, questions of globalization and uneven development, politics of diversity and spaces of resistance, or geopolitical questions related to displacement–the environmental and human interconnectedness that geographers deeply ponder has always fueled my own imagination.

how to adapt to new circumstances and appreciate other cultures' diverse approaches to daily life.

Achieving her postsecondary education was a long process with a few interruptions along the way. She completed her first bachelor's degree in geography at Elphinstone College in Mumbai and graduated with the highest marks in geography at the University of Bombay (now the University of Mumbai). She was admitted, with multiple scholarships, to graduate school. After a year, however, she moved to Toronto, this time with her spouse. After arriving in Canada, she worked for a few years as a cartographer in the days before computers, when Rotring pens and manual typesetters were used. She wanted to complete her graduate education, but now she had to factor in two children. To fulfill admission requirements, she completed additional undergraduate courses, which included her first course in geographic information systems (GIS). In 1997 she completed her master's and in 2002 her PhD at the University of Toronto, all while working part time, earning a limited income and taking care of a family.

On every step of her journey, she was encouraged and supported by her spouse and the community around her. "A child-care center with wonderful caring staff at the university allowed me to pursue my studies with confidence while a network of women provided a community of support in my neighborhood. The neighborhood was home to primarily migrant working-class communities with extended families, including many grandmothers! As childcare was expensive, an informal and unspoken culture of caring and watching each other's children was the norm. Doors remained open, and children ventured into and out of each other's homes. Food was shared, carpooling [was] regular, homework space [was] open, and celebrations were galore! Writing a dissertation (on the neoliberalism of education and politics of school closures in Toronto) in the midst of this made it all seem so much easier!"

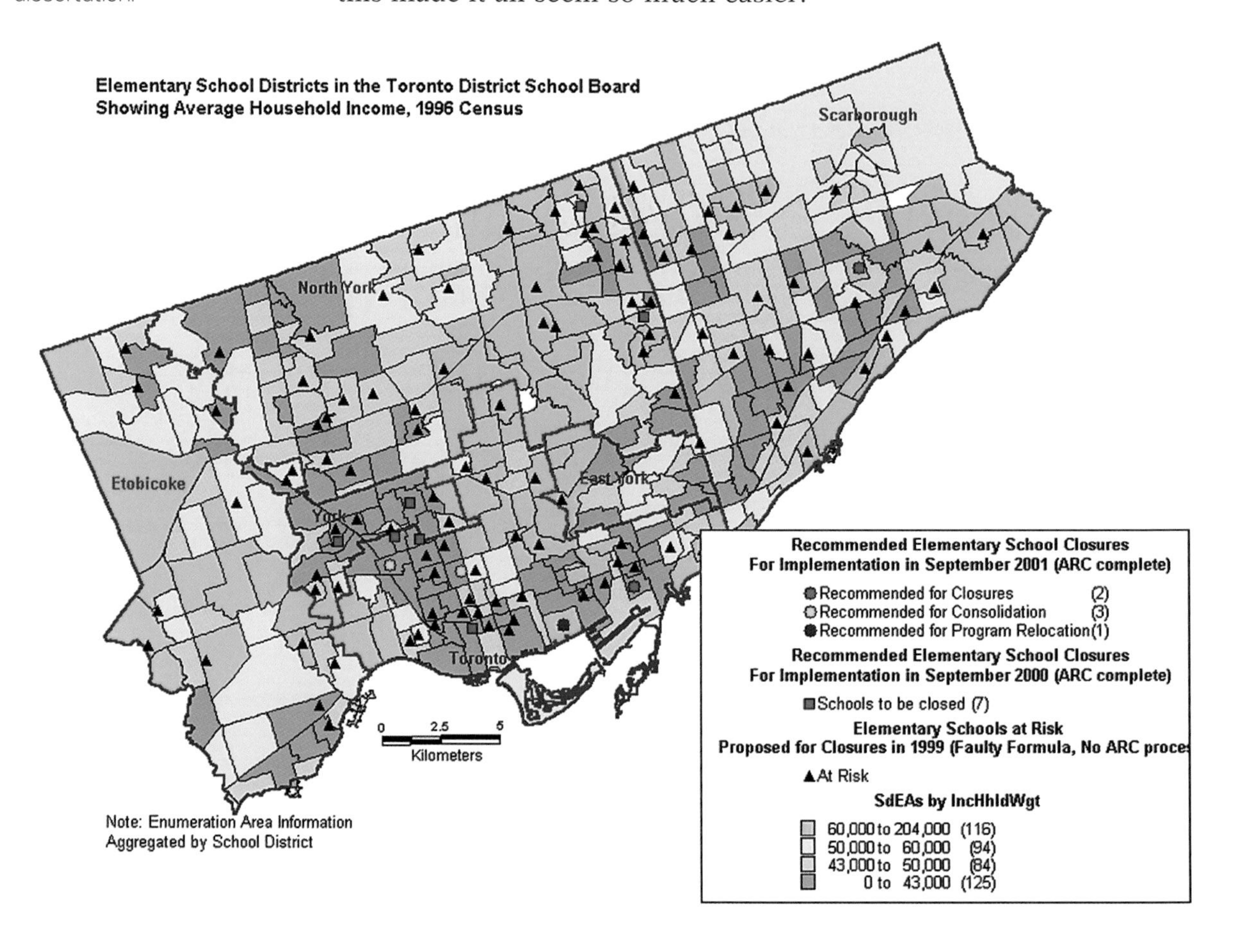

Ranu created this map showing school closures in Toronto as part of her dissertation.

GIS is not just a descriptive mapping exercise leading to the production of visually appealing and powerful maps, but a deeply heuristic process that requires reflexive and theoretical consideration. I thus try to impart [to] my students the importance of using a "critical eye" when reading and analyzing maps; and to reflect carefully on the underlying historical/economic spatial processes and political/social structures that lead to such outcomes and [their] broader implications, whereby questions of inequity, stigmatization, and oppression are not further reproduced through the process of careless mapping and interpretation.

The faith that her PhD supervisor, Professor John Miron from the University of Toronto, and other mentors placed in her also gave Ranu strength and supported her through the challenging years. "Prof. Miron was a pioneer in the field of GIS in human geography (when GIS was primarily restricted to physical geography) and spatial statistics in location theory. He taught me not only the methods and means of empirical inquiry using GIS in the social sciences, but also the many challenges of GIS mapping and cartographic representation, especially in human geography," she says.

One aspect he emphasized, which is often ignored with the large proliferation of data and ease of mapping, is questions of how the process of mapping needs critical interrogation both technically and philosophically, as there are issues of ethics and privacy that need to be carefully considered.

Currently associate professor in the Department of Geography at York University, Ranu's research and teaching interests relate to the geographies of marginality, diversity, and social justice in cities; power, space, and activism; critical geographies of education; social sustainability and the meaning of public space as it relates to migrants; and the provision of infrastructure for marginal groups in suburban regions. She explores these various themes through her research, activism, and writing projects.

Recently, she embarked on a five-year Social Science and Humanities Research Council of Canada-funded project titled *Subalterity, Education, and Welfare Cities* that traces the historical and geopolitical impacts of conflict and displacement on cities and schools in Havana, Cuba; Toronto, Canada; and Kolkata, India. Ranu has served on executive boards at her university for many years, including the Centre for Refugee Studies (CRS), the City Institute at York University (CITY), and the York Centre for Education and Community (YCEC), and currently serves on the York Centre for Asian Research (YCAR) Executive Committee.

Ranu visits an old schoolhouse at Biran Farm, Cuba.

Ranu, who says she enjoys the privilege of working with many committed and talented colleagues and students at York University, is deeply inspired by the communities she has observed in Cuba and India. She has met diverse groups working passionately to build a socially just society, maintaining their high ideals and standards as they face constant challenges on a daily basis. For instance, in Cuba, despite over five decades of an economic blockade, the educational and health-care systems are world renowned because of the commitment and solidarity of its people. In Kolkata, teachers in municipal schools speak of their daily challenges yet remain determined to bring children who are street vendors into their classrooms, even if for only a few hours. In Toronto, local migrant communities organize to build subaltern cosmopolitan communities of support sharing common experiences of exile and pain.

The field, she notes, is a place of learning/unlearning, activism, and critical reflection. Using a method termed *spatial phronesis* (including work in GIS), which involves practical wisdom gained from ethnographic inquiry, reflexive observation, and active participation, she spends time in libraries and archives, collecting data from diverse sources. Ranu also works alongside educators in the classroom and participates in social and political events. One of her favorite experiences was in Kumirmari, a remote island in the Sunderbans at the confluence of the Ganges and Brahmaputra Rivers and at the edge of the vast Sunderban forest, home of the Royal Bengal Tiger and other wildlife. Ranu's mother had also spent many childhood years in Kumirmari prior to India's independence. Over time the island has experienced many challenges—from flooding disasters, constant migration flows, and political unrest to abject poverty and developmental conundrums. For Ranu, working with

Ranu with schoolchildren on the Sunderban island of Kumirmari, India.

the educators, children, and residents on the island was a transforming experience—personally, intellectually, and politically.

Ranu encourages young women to follow their own inspirations, to persevere and constantly challenge themselves, and to not let anyone disillusion or discourage them. She advises them to be astutely aware of historical and structural conditions that lead to disadvantage or privilege. She advises women to find allies and build networks of support, and to engage in "slow scholarship"—a feminist movement advocating for a slower pace and supportive scholarship allowing the time to be reflexive while living, thinking, and experiencing academic life more holistically. And she encourages them to take time to unwind and have fun.

Ranu will continue to explore questions of equity and social justice, rationality and power relations in planning decisions, and trends in geopolitical displacement resulting from violence and conflict. She wants to raise awareness of systemic impediments such as race, class, and gender divides that continue unabated and the resistance that follows from within. In the future, she envisions further engaging in community mapping with GIS in refugee camps and remote communities and schools as a critical political and empowering tool.

We need to build a society of support systems and international solidarity that takes care of its most vulnerable populations, especially those who have been forced to the margins. In a world driven by market logic, women often continue to bear the brunt of additional societal challenges related to social reproduction inequities and precarious working conditions, violence, and exclusion. *

Deirdre Bishop

Keeping track of the census

Deirdre's love of the arts has transformed into a career in the GIS industry.

DEIRDRE DALPIAZ BISHOP BELIEVES THAT HER LOVE OF GEOGRAPHIC information science started with the arts. Born in Binghamton, New York, Deirdre had a father who was an architect and a mother who was an art teacher. Her family liked to travel to museums, architectural sites of interest, and places of natural beauty, which helped her gain an appreciation for art and design and fueled her interest in the study of cities. Years of watching her parents teach art to students, design buildings, paint, create mosaics, and tend to a lush garden gave her an appreciation for all things beautiful. Although she did not inherit their level of skill or passion for creativity, she fell in love with color and form. When her career in the GIS industry started to take shape, it was a different kind of design that enabled her to build color and shape into her everyday world.

Deirdre started her college education at Lehigh University in Bethlehem, Pennsylvania, with the hope of becoming an accountant. After three years of majoring in accounting and minoring in urban studies, her adviser and mentor, Professor David Amidon, recognized that her passion was in urban studies and encouraged her to pursue an internship in city planning over the summer.

She spent the next three months interning in Binghamton, New York's Department of Planning and Economic Development. She still has her journal detailing her daily experiences and accomplishments. It changed her perspective, leading to what she feels was one of the best decisions of her life. She returned to Lehigh as a senior and changed her major to urban studies.

Throughout high school and college, Deirdre participated on the cross-country and track teams. Not only did the strenuous practice and success at competition help build her self-esteem, but her selection as captain of these teams provided her an opportunity to demonstrate her leadership skills and motivate others to succeed—an experience she is certain helped prepare her to lead in the workforce later in her career.

Deirdre's first job was at the Southern Tier East Regional Planning Development Board, in Binghamton, where she learned how to translate the US Census Bureau's Topologically Integrated Geographic Encoding and Referencing (TIGER) system files into a format that could be used in GIS. This knowledge qualified her to apply for a job as a geographic specialist in the Census Bureau's New York Regional Office (NYRO) in Manhattan, the city of her dreams.

At NYRO, she managed all Census 2000 geographic operations for nine counties in New York, including New York City, and 10 counties in northern New Jersey. She managed the Local Update of Census Addresses Operation by working with tribal, state, and local government officials to ensure the accuracy of the Census Bureau's national address list. While there, she formed a strong working relationship and friendship with Dr. Joseph Salvo, director of the New

York City Department of City Planning, Population Division. To this day, Joe continues to serve as a mentor and friend.

She also developed NYRO GIS, a tool for visualizing and sharing demographic data that could be used by all managers in the New York office, both geographers and nongeographers. She was twice honored with the Census Bureau's Bronze Medal for Management of geographic operations for Census 2000 in the NYRO and for leadership in response to the September 11 attacks on New York City.

On September 11, 2001, Deirdre and her husband were walking to work across Bleecker Street in Manhattan when they heard a tremendous roar above their heads, and then a crash that sounded like a garbage truck dropping a large container. When they turned to their left and looked downtown, they could see the outline of the first plane that crashed into the World Trade Center (WTC). They naively said goodbye to each other and headed to their respective offices. By the time Deirdre arrived at her office on Varick Street, the second plane had crashed into the other WTC tower. Within an hour or so, the whole office was excused from work and instructed to go home, but for many of her colleagues, crossing the Hudson River to New Jersey or getting across the bridges to Brooklyn and Queens was not an option. Instead, about 20 of them headed to her one-bedroom apartment on Houston Street. They turned on the TV and learned that Flight 93 had crashed in Shanksville, Pennsylvania. Two of their colleagues were on that flight. Between the tragic events of that day and the loss of two leaders in NYRO, her office struggled to recover. It was during this time that she learned an important lesson about leading people:

It's critical to care for your employees as people—not just for the work that needs to get done. If your colleagues and staff are not well, it will be nearly impossible to do a quality job. I have carried this lesson with me throughout my career.

Deirdre working Census 2000 in the New York regional office, circa 1999-2000.

For women who are starting off on their careers, she advises volunteering for tasks outside the direct line of their responsibilities. One of the best experiences she had in NYRO was leading the lunch-time exercise program. Another valuable lesson she learned while working in the Census Bureau's Budget Division was to help with whatever was needed, demonstrating a will to go the extra mile, including stuffing binders to meet a deadline.

Deirdre received her master of public administration from New York University and moved from Manhattan to Washington, DC. There, she became deputy chief of the Redistricting Data Office and learned the importance of finding a mentor. Chief Cathy McCully, whom she reported to, demonstrated to Deirdre how to be a strong leader, woman, and mother, which was especially significant to Deirdre because her daughter, Fiona, was born during this time. Cathy also pushed Deirdre to apply for the Executive Leadership Development Program at the Census Bureau.

Deirdre uses GIS for her work on Census 2000 in the New York regional office, circa 1999-2000.

When she was in the Census Bureau's Geography Division as geographic adviser/deputy division chief, Tim Trainor encouraged her to apply for the Senior Executive Service Candidate Development Program (SESCDP). Another mentor in her career, Tim was conscious of recognizing and promoting women. Deirdre remembers:

"When I was serving as his adviser, from 2009 to 2014, I did not quite understand his intent. I felt as though I always had equal opportunity and recognition. Looking back, I think this was because I received a tremendous amount of positive reinforcement from family, teachers, and coaches. They encouraged me to excel in school and sports, which helped me to feel I was on a level playing field. In addition, working with people was fun for me—whether they were young or old, black or white, outspoken or shy. I enjoyed hard work and solving problems, especially when that involved a team."

But she noticed a difference in the workforce.

It was not until I joined the Senior Executive Service within the federal government that I began to notice the disparity. While the number of women involved with census taking and geographic information science is high, the number of women leaders is not. This is also true of leaders within the international organizations that I serve.

During her participation in the SESCDP, Deirdre pursued an assignment at the National Opinion Research Center (NORC) and designed a new strategic planning process for the Department of Commerce. She was then promoted to serve as chief of the Decennial Census Management Division, where she designed the census of 2020 and released the 2020 Census Operational Plan three years early, in 2017, for which she was awarded the Census Bureau's Bronze Medal.

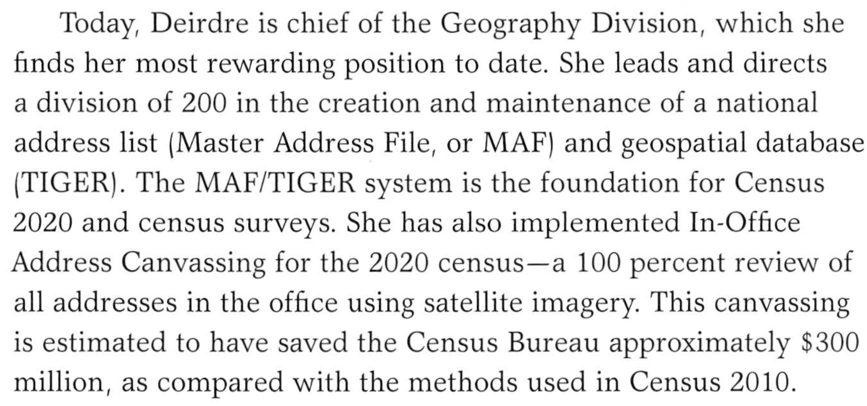

Today, Deirdre is chief of the Geography Division, which she finds her most rewarding position to date. She leads and directs a division of 200 in the creation and maintenance of a national address list (Master Address File, or MAF) and geospatial database (TIGER). The MAF/TIGER system is the foundation for Census 2020 and census surveys. She has also implemented In-Office Address Canvassing for the 2020 census—a 100 percent review of all addresses in the office using satellite imagery. This canvassing is estimated to have saved the Census Bureau approximately $300 million, as compared with the methods used in Census 2010.

She also serves as head of the US delegation for the United Nations Committee of Experts on Global Geospatial Information Management (UN-GGIM) and as president of the US National Section of the Pan American Institute of Geography and History (PAIGH).

Deirdre is head of the US delegation for the United Nations Committee of Experts on Global Geospatial Information Management.

Deirdre believes that women must support, mentor, and lobby for each other. Developing self-esteem and confidence in young women is especially important, as she sees with her own teenage daughter and with the new professional women in her office. She strongly recommends that young girls and women take a diverse course load in school with a focus on multiple languages if possible; participate in science fairs, math competitions, and organized sports; and travel, both domestically and internationally. This combination helps build knowledge, interest, and self-esteem, she says.

Overall, Deirdre wishes that young women would not be so hard on themselves. As she says:

You are going to make mistakes. If you work hard on a day-to-day basis and show that you are committed to the team, your boss is going to understand if you goof up. More likely than not, they will help you through it, and then forget about it the next week.

She recently attended a United Nations Economic Commission for Europe Meeting of the Group of Experts on Population and Housing Censuses. During the break, she picked up a copy of *UN Special* magazine (no. 782, September 2018). The centerfold of the magazine contained a diagram titled "The Cognitive Process and Our Ability to Learn" and a highlighted box about STEAM—science, technology, engineering, the arts, and math. She was pleased to see this focus because she believes it has come full circle—her interest in GIS started with the arts, and the arts have continued to apply to her career in STEM and with GIS. ✸

Paulette Brown-Hinds

Bringing community awareness to life

Paulette is committed to using GIS in education, publishing, and community service.

DR. PAULETTE BROWN-HINDS HAS A DEEP PASSION ABOUT ALL ASPECTS OF building community: educating, informing, enlightening, engaging, and inspiring the people around her. She has nurtured this passion over time, through education, publishing, and community service. Her latest project is Mapping Black California, a community mapping project that uses GIS to better understand data on African Americans in the Golden State. It's also designed to introduce the power of GIS to the African American community, mainly through partnerships with nonprofit organizations working with young people.

Paulette comes from a family of committed community servants. Her father, Hardy, and her paternal grandparents were sharecroppers in rural North Carolina. She can trace that part of her family back to slavery and to the McDaniel family of Jones County. For generations, the Browns not only lived and worked in that region, but they were also members of civil rights and other civic organizations that fought for equal rights for African Americans. Her grandfather, Floyd, risked his life to register blacks to vote shortly after the Voting Rights Act was passed. Her grandmother, Essie, raised nine children while working to help provide for the family. Their home became the center of the black community. Weary passersby knew they could always stop to share a batch of biscuits or chat about the important political issues of the day. Their porch was a safe and sacred space, where people gathered for information and encouragement.

Paulette's mother, Cheryl, grew up in a household that valued service to others above the self. Paulette's maternal Grandma Melba was known to take in anyone who needed shelter. Paulette's grandmother had three children of her own, but when her sister died unexpectedly, she took in her five nieces and nephews, as well as another niece, for a total of nine children. She was also civic minded and encouraged her children to be community problem solvers. Paulette's maternal grandfather Marvin's income tax business, now run by family members 60 years later, became an anchor small business in its south Los Angeles neighborhood. He also engaged the community's youths by sponsoring sports teams and community bowling leagues. He even placed a table and a chess set in the front window of his office building, encouraging neighborhood kids to come inside and learn how to play.

Paulette's childhood was built on that foundation. Her family dinner conversations were about church, scouting, local politics, and school activities. Her father was elected to the city school board, and they often discussed equitable solutions to education problems that plagued the most vulnerable populations. During campaign season, they talked voter registration and voter education. She was offered her first job, as a phone bank caller for political campaigns, when she was 15.

Things were no different for her at school. One of her first memories in elementary school was when she was asked to join

"the lunch bunch"—a group of students who were selected to host special guests and dignitaries on campus. She didn't know it then, but it was an introduction to work she'd find rewarding as a community servant. She felt special to be entrusted with such a responsibility—they were given plastic smocks to wear as "uniforms" and had the freedom to miss class.

During those years, she credits a string of teachers who made her believe she was a valued contributor: Ms. Box, Ms. Fountain, Mr. Kissinger, Ms. King, and, of course, the teacher that was the hardest on her, Ms. Jordan. They had high expectations of their students, and Paulette worked hard to meet them.

In middle school, her PE teacher, Ms. Dyer, encouraged Paulette's idea to start Big Sisters, a group mentoring the younger girls who were struggling to find their way. Big Sisters was designed to identify the girls in need, confidentially share their problem with female upperclassmen, and devise a strategy to assist as a community of supporters. In high school and college, she continued to develop programs and use her connections to improve the campus community. It wasn't until well into adulthood that her parents confessed to her that a pair of her middle school teachers told them during a parent-teacher conference that she wasn't as smart as they thought she was. "My parents waited until I was 30 and had earned my doctorate degree to let me know."

As an undergraduate English major at California State University, San Bernardino, Paulette found herself immersed in the world of literature of the African diaspora. One of her professors, Dr. Misenhelder, had just returned from a year-long Fulbright Fellowship in Botswana, and she was eager to teach her students what she learned about South African culture and literature. That became Paulette's introduction to hearing the voices of the disenfranchised that resonated with her own experiences. The more she read about the freedom fighters of the apartheid system, like Nelson Mandela and Steven Biko, the more anxious she became to engage in activism. As an undergraduate student, she started to attend rallies that called for state colleges and universities to divest from South Africa.

She then entered the University of California, Riverside (UCR), doctoral program in English literature. She expanded her studies to Afro-Caribbean women writers and was fortunate to find supportive professors. Emory Elliott, one of the most respected American literature scholars in the world, and Sterling Stuckey, one of the most respected scholars focused on slave culture and black nationalism, became her mentors.

Josiah Bruny, *front row center,* founder and CEO of Music Changing Lives, speaks to Paulette's class on arts and the community at UCR. Paulette is standing, on the right.

One of her biggest hurdles was having a child while she was in the middle of graduate school. She remembers sitting for her master's degree exams while eight months' pregnant, wondering if she could manage both graduate school and motherhood. One of her fellow classmates mocked her because she was pregnant. "Doesn't she know there's something called birth control?" he said to one of her friends. And later, she remembers not getting a job because she commented to a female professor that she and her husband, Rickerby, didn't want to put their son, Alexander, in day care too early. "Well, I guess I'm a bad mom," the professor said in response, as if Paulette had made an indictment on her parenting choices.

When she didn't get the job, she was crushed. She will never forget Professor Elliott's words of encouragement: "I don't know how to get you to see yourself the way that I see you," he said. She applied for more positions, landed at the University of Cincinnati's Department of English and Comparative Literature, and was added to the faculty of African American Studies in a joint appointment. During her brief time in academia as a full-time professor, she traveled the world presenting research, published a variety of academic essays, was awarded grants in service learning, designed innovative courses incorporating literature and community service, and won a teaching award. But with all her success, she knew that it was not where she was supposed to be. She longed to return home to the community that nurtured her all those years.

Paulette returned to her hometown and joined the family business, first as executive director of the educational nonprofit Black Voice Foundation and then as managing partner of Brown Publishing Company, the firm her parents created in 1980 to manage the *Black Voice News*, a weekly advocacy newspaper founded by UCR students in 1972. Her parents turned the newspaper into a nationally recognized, award-winning weekly publication, and published in the advocacy and activist tradition of the Black Press of America, pointing out discrimination and fighting for social justice.

In 2012, Paulette took over as publisher of the newspaper as well, and rebranded it as Voice Media Ventures.

Paulette continues to publish the *Black Voice News* (under the banner VOICE) and operates a strategic communications firm that specializes in community engagement in California, a multimedia firm that produces films and video content, and an educational and philanthropic division that partners with school districts to teach empathy and community building through cultural study tours. She has always been inspired by black women news publishers who came before her: her mother, who was a relentless advocate for the voiceless and disenfranchised; Ida B. Wells, whose fiery editorials against lynching in her community of Memphis, Tennessee, were so incendiary that she was forced to live in exile; and Mary Ann Shadd Cary, who was the first woman to publish a newspaper in North America.

Paulette's interest in GIS started after she and her staff used Esri® Story Maps software to map the route of an Underground Railroad study tour. The map traced the route of an eight-day trip from Kentucky to Canada using photos, site descriptions, and historical information. A few months later, the same team created a story map for a California Coastal Commission grant tracing the history of segregated beaches that dotted California's coastline in the past, "The History of California's Black Beaches: Segregation by the Sea," at https://bit.ly/2PqxWVv.

Through her experimentation with story maps, Paulette saw the potential in continuing the historical advocacy of the black press with the data visualization capabilities of geospatial technology. Paulette called the project Mapping Black California, although at the time she had no idea what it would entail.

Here was an opportunity to take the newspaper's 45 years of valuable community data and create interactive, informative, and visually stimulating stories. These stories can become assets to policy- and decision-makers, business and foundation leaders, educators, and advocates.

She expanded on her ideas after being invited to the Esri campus in Redlands, California, to attend an annual presentation by high school students using ArcGIS® software to understand complex problems in their communities and explore possible solutions. In coordination with musician/activist will.i.am's "i am angel foundation," the program has provided the software to Roosevelt STEM Academy students in the Boyle Heights community of Los Angeles since 2013. Paulette found inspiration in the work of the high school students and wanted to include the same type of learning opportunities in her community mapping project.

Paulette, *left*; Esri president Jack Dangermond; and Paulette's mother, former California state Assemblywoman Cheryl R. Brown, *right*.

By using a community mapping framework and partnering with schools, universities, community-based organizations, and businesses, Mapping Black California has been able not only to tell the story of black California, but also to better interpret data and history. Through this community mapping project, they are also addressing the growing national concern over the lack of diversity in the technology industry and helping to grow the population of African American geospatial experts.

Paulette's aspiration for prompting more African American GIS analysts also resonates with others. She explored the union of the *Black Voice News'* history and the potential of GIS with a group of key stakeholders interested in participating in the initiative. Since that meeting, two nonprofit organizations that introduce African American students to the technology field have designed and added GIS modules to their curriculum.

Paulette's future work will continue to connect media, technology, and philanthropy for the public good. She was recently elected to serve on the James Irvine Foundation Board of Directors, and through that entity she has funded some of the nonprofits that work with African American youths (especially teenage girls) and that have added GIS to their programs. She is the founding instructor of a UCR course for creatives called Arts Management and the Community. The theme is "The Start-up of You" and offers students a chance to connect with the surrounding community as they begin to think about their professional lives as working artists.

Paulette is a believer in the West African concept of "Sankofa"—the Ghanaian proverb in the Akan language of Ghana, which translates to "reach back and get it." The proverb is represented by the Adinkra symbol of a bird with its head turned backward and an egg on its back. Paulette's message to young women gives voice to this proverb:

Tata Donets/Shutterstock

Young women are capable, they are powerful, and they are smart. They can do anything they want if they are courageous and creative. And the only limitations they face are the ones they set for themselves. ✱

Molly Burhans

Turning landownership into land stewardship

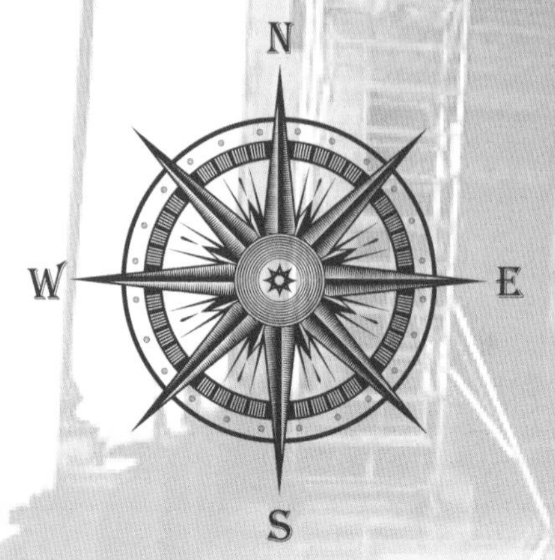

MOLLY BURHANS COMBINES HER PASSION FOR PEOPLE AND THE PLANET with her abilities and experience in design thinking, business development, and scientific research. She has found inspiration in spiritual leaders, artists, poets, writers, philosophers, scientists, and friends from all walks of life. Three years ago, she founded GoodLands, a social enterprise with the mission of helping clients use their land for good. GoodLands is helping the Catholic Church understand ways in which its extensive real estate holdings can be used to support environmental health, social justice, and financial viability and growth. As one of the largest landholders in the world, the Catholic Church could have an incredible impact on the climate, and thereby on all of Earth's living creatures.

Molly has proved that reinvigorating a mapmaking tradition with a modern GIS has the potential to benefit not only the church, but the lives of people all over the world. In the summer of 2018, GoodLands received permission to test-run the establishment of a cartography institute in the Vatican. If Molly and her GoodLands team are given permission to establish a permanent cartography institute, it will be the first female-founded and female-led institute in the Vatican. Molly believes in what she is doing and says she has accomplished her goals because of her abilities to connect and unite the natural world and technology.

Molly grew up roaming the halls of the State University of New York at Buffalo's Computer Science Department, where her mother was completing her PhD, and playing between the benches of her father's research lab in a cancer center. Her early technology experiences, which included sitting in the back of her mother's classes, taught her that there were many ways to use computers besides basic games and creating documents. When young, she was introduced to Adobe® Macromedia MX software, which led her into the world of coding, Dreamweaver®, Photoshop®, and—an early favorite—Flash®. She was mostly self-taught, and design programs provided her with endless hours of amusement. As she developed animations with her own graphics, she didn't realize she was acquiring the skills and knowledge that would one day be important to her career.

When she was not absorbed in her computer screen, she spent time with her family, hiking and backpacking in Vermont, California, and Wyoming. Her grandfather seemed to know the names of nearly every plant and always made time to stop, smell, and identify the flowers. These experiences created within her a deep interest in, and respect for, all aspects of the natural world. Exploring the outdoors with her scientist parents profoundly shaped her ability to use language, art, and computational design to describe nature. It is no coincidence that John Muir's appreciation of the natural world and his actions to protect it inspired her in later years.

Molly has always had a unique approach to learning. She needed to see the big picture and visualize both the parts and the whole. This led her away from the sciences through high school and college and into philosophy and the fine arts. When she was 14, she drew a scientific figure for a paper her father published. This start led to additional jobs in scientific illustration, including the creation of figures, conference posters, abstract and book covers, and other visual media. These early career experiences guided her to integrate art and science in her mind.

When she studied at Canisius College, many of the Jesuit priests became important mentors and friends. Her spiritual director, Father Thomas Colgan, SJ (Society of Jesus), counseled her through Saint Ignatius's spiritual exercises and discernment of spirits for over a year. This process helped form her faith and values as she returned to the Catholic Church, at a time in life when people often leave their faith traditions. She may have had an inkling of her future, however; Saint Hildegard von Bingen, a Catholic polymath, mystic, and abbey foundress, who counseled high-ranking Catholic leaders in her time, including the pope, inspired Molly so much that she chose her as her confirmation saint. Molly feels her life would not be as rich without her continued learning in theology.

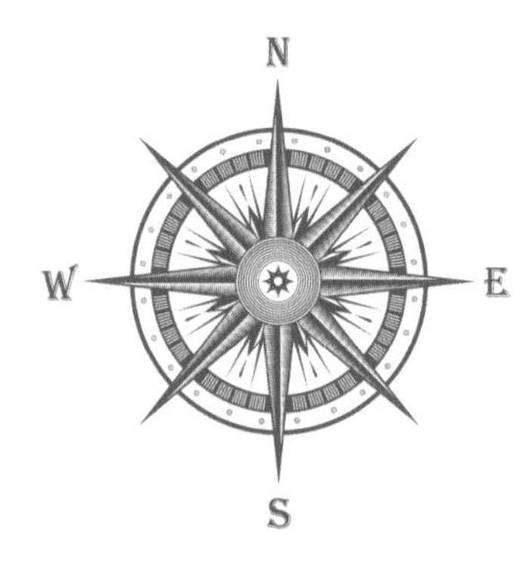

Molly made her first map in graduate school. Her class was required to go into the field with surveying equipment and create a map to scale, by hand, on layers of tracing paper. The day that she opened the GIS software ArcMap was one of the best days of her life, as if someone had taken her mind and placed it directly in a software program. Molly received her master of science from the Conway School in western Massachusetts. The school provided her with a learning environment that integrated disciplines, enabling her to flourish. Paul Hellmund, greenway design expert and school president at the time, became a close mentor.

During graduate school, Molly met Jill Ker Conway, whom Molly calls an amazing, strong, feminist, brilliant, faithful woman. Jill was an author, Smith College's first woman president, and a visiting professor at the Massachusetts Institute of Technology. When they first met, Molly was scared to tell her or anyone about her ideas for

Catholic land because they seemed far too ambitious. Once Molly told her, Jill looked at her and laughed: "Well, that's a lovely first 10 years to your career!" She became GoodLands' first donor and guided Molly through its early development. She passed away in the summer of 2018, but Jill's generous spirit and optimism continue to inspire Molly every day.

She decided that after graduate school, she wanted to join the "Nature Conservancy" of the Catholic world, or the cartography or planning department of the Vatican. She was shocked to learn that the organization that oversees the largest networks of health care, education, and aid in the world had neither a center nor NGO partners addressing conservation and land use. Equally surprising was learning that digital land record systems are rarely used by Catholic communities despite the quantity of landholdings and their rich land-tithing and donation systems. Molly founded GoodLands to fill this void.

The idea for GoodLands originally emerged in her late teens. Molly regularly attended retreats at a Benedictine monastery, where she began seriously considering becoming a nun. At the same time, she cofounded her first company, GroOperative, a worker-owned aquaculture and agricultural enterprise in Buffalo, New York. At the time, she was studying at Canisius College, developing scientific media, working in a biology laboratory, and independently studying mathematics, early cybernetics, and design. But at the monastery, she was enormously impressed with the good that the sisters were achieving—getting rival gangs to reconcile and lifting everyone's lives in an entire neighborhood. She was also concerned about the future of their properties and noticed opportunities for better land and property management that could help support their mission.

Early in GoodLands' development, Molly found herself on the third loggia of the Vatican Apostolic Palace in a hallway lined with some of the most spectacular maps she had ever seen. She asked representatives from the Holy See Secretariat of State where their cartography department was and learned that they had not overseen the creation of an updated map since the ones painted centuries ago. This led Molly to conceive of the most important and ambitious GIS project of her career: mapping the Catholic Church.

If faith the size of a mustard seed can give us the strength to move mountains, I am confident that it can enable us to map them and contribute to their conservation.

While Molly was in graduate school, GIS opened her world to computational landscape architecture and the possibilities for planning large sets of landholdings. She directed the development of the first comprehensive global data-based maps of the Catholic Church. These maps are also the first data-based global maps of any major world religion drawn along geo-religious boundaries. The development of the datasets for the Catholic Church required a

The third loggia of the Vatican Apostolic Palace.

large team that included GoodLands' interns and contractors as well as individuals from Esri's cartography, applications prototype, and professional services teams. For the first time in history, these maps illustrate the lands and people of the Catholic Church covering the surface of the earth.

Molly working late into the night at Esri's Redlands, California, campus.

These maps were exhibited as part of the Vatican Arts and Technology Council meeting in December 2016 in Casina Pio IV, which houses the Pontifical Academy of Sciences. The council's project is driving discussions concerning canon law in the digital age, church management structures, and global responses to climate change. In the summer of 2018, GoodLands received papal permission to test-run the establishment of a cartography institute in the Vatican.

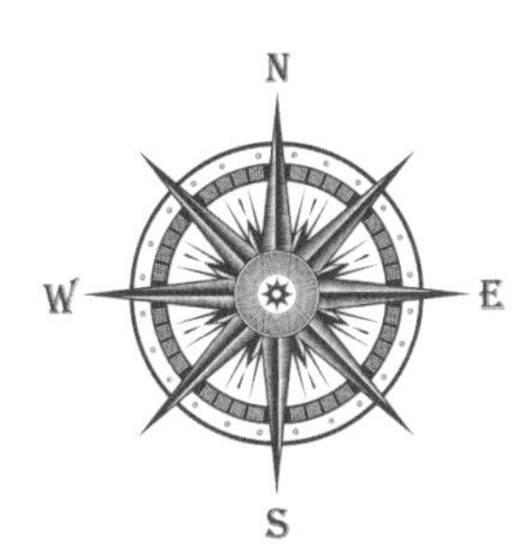

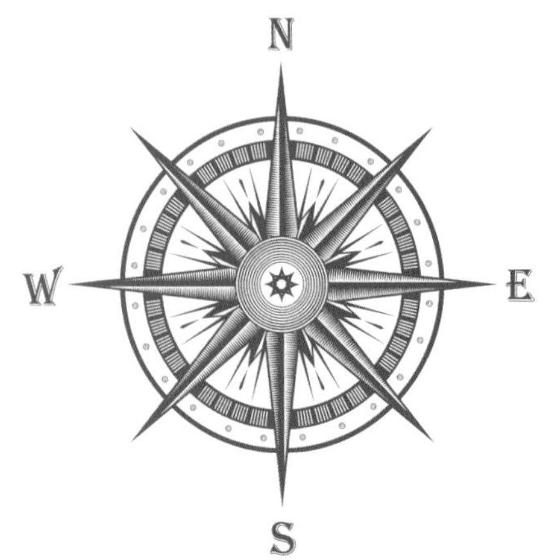

Maps in Casina Pio IV II, the Vatican.

Molly and her team have continued to update these maps and enrich and analyze the information to develop hundreds of additional maps and applications from this foundational dataset. They have mapped priest shortages, sisters around the world, various Catholic populations, and clerical abuse. They have connected climate modeling to the system and are in the process of working with Georgetown University to tie in predictive migration modeling. This project is enabling the development of an early warning system that can show which dioceses are at elevated risk for famine, migration, disease, and conflict. Further, the project reveals key information about church leadership, potential land utilization, and locations where service sectors must get involved to respond to emerging humanitarian and climate disasters.

Molly's goal is for GoodLands to make major contributions toward helping mitigate climate change and its consequences, including climate-driven migration, over the coming decades. She wants to expand the company's operations to partner with other religious groups, individuals, and companies that own substantial distributed networks of land. Molly feels that it is essential that, in the coming years, her services, educational work, and marketing outreach help the broadest community of property holders transform their sense of landownership into a sense of land stewardship through a deepening understanding of the moral dimension of land use and management.

Starting a new organization has unique challenges not only because of the gender gap in financial investment, but the entire climate of philanthropy, Molly says. Foundations are calling for financially self-sustaining nonprofit business models and scalability. Nonprofits, particularly those just starting out, must be competitive, impactful, and able to generate returns while using limited capital to be scalable.

Molly's main inspiration comes from the top. After Pope Francis's encyclical papal letter *Laudato Si'* was released, she founded her company. Pope Francis's ongoing concern for the planet and its people serves as a foundation for her goals. Molly's principal mentor of the last few years, Rosanne Haggerty, whose work with the homeless greatly inspired Molly, opened her home in Hartford, Connecticut, to Molly for the crucial beginning years of getting GoodLands off the ground. Rosanne is an adviser to GoodLands, a business mentor, and a friend whose counsel and generosity were crucial to GoodLands' development, Molly says. She is also inspired by Carl Steinitz, Stephen Ervin, Dana Tomlin, Darrel Morrison, Bill Mollison, Arancha Muñoz-Criado, Donella Meadows, Bill Drayton, Father Robert Maloney, and Buckminster Fuller among others. So many people have helped and inspired her on the journey, Molly says.

Also, she notes that there is much more at stake when working for the common good: if, collectively, the varied approaches to mitigating climate change do not scale sufficiently to reach critical impact, it won't be the market that crashes, but the entire global ecosystem. That is a heavy weight to bear but one that generates greater momentum.

I see social entrepreneurs, hand in hand with impact investors, as attempting to hack capitalism, to mature it from its current, ethically blind state towards a more humane, mature state—where mission and margin are intertwined. Developing a social enterprise is immensely challenging but equally exhilarating, requiring enormous creativity and ferocious energy.

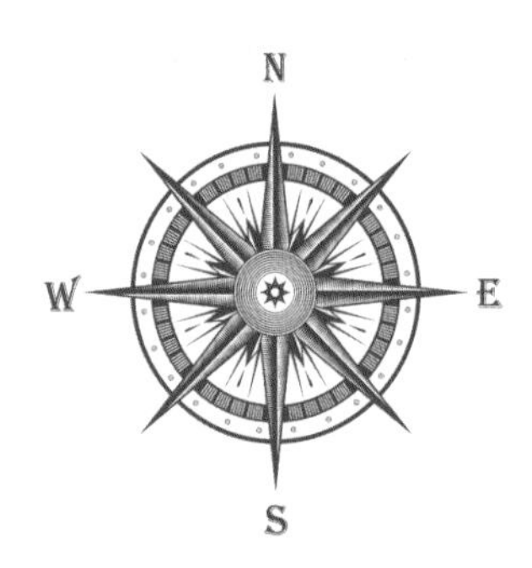

In 2018, Molly, seen here with Pope Francis, received papal permission to test-run a cartography institute in the Vatican.

Molly is a force to be reckoned with: she is incredibly persistent and willing to put in an enormous amount of work to accomplish her goals. To get where she wants to be, she is willing to make great personal sacrifices. Molly recognizes, however, that although this quality is a strength, it can also be a weakness. For Molly, finding a good balance is an ongoing challenge. For better or worse, she speaks her mind and is true to herself, and this authenticity has made a huge difference in helping her share her vision, inspire others, and connect with genuine, caring, talented people.

Molly's advice to young women:

Let yourself be drawn to what you love about STEM, whether inside or outside formal educational and research settings. There is often a one-size-fits-all view of education–if you do not fit within this view, you can still work in STEM.

Her education was not standard. It involved adventures, books, online materials, lab work, occasional courses, nature, and hours discussing questions with peers. She encourages young women to be strong in what they know—and to recognize that there will be roadblocks, including frustrating experiences of bias.

Molly has been fortunate to be welcomed into networks full of enthusiastic and respected researchers, educators, and entrepreneurs. She feels that it is immensely fulfilling and enjoyable if you can align your career and life in such a way that you are surrounded by people who love what they do, and the best way to achieve this is to do what you love.

Never stop being in wonder of your work. There is infinite wonder and beauty in a single blade of grass, an interesting rock formation, and the face of another human being. A million scientists, artists, and spiritual leaders could never fully express this, but we are given a lifetime to attempt it. Exuberantly embrace the brief but magnificent opportunity to do so. ✸

Kate Chapman

Open to helping others benefit from technology

Kate spreads her love of geography to nonprofits.

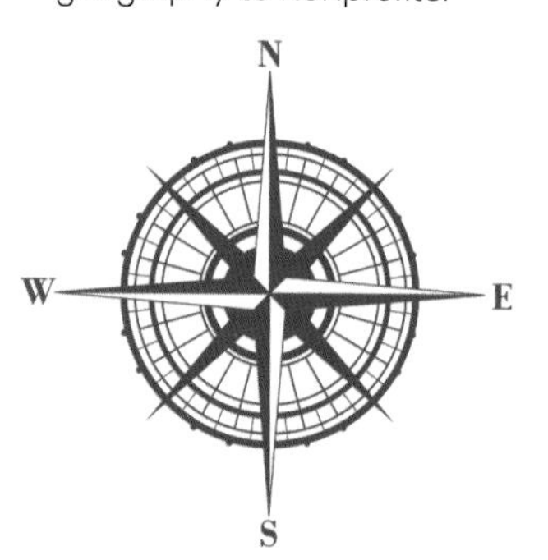

IF KATE CHAPMAN HAD NOT BEEN ALLERGIC TO PESTICIDES, SHE MIGHT have had a different future. Her first GIS job at college was making mosquito treatment maps when the West Nile virus was making its way across the United States. She was originally hired to apply a pesticide, but when she had an allergic reaction on the second day, she switched to making maps. This change of events inspired her to switch her major to geography and ultimately start her own company to advocate for geography-based solutions.

After being pulled off pesticide duty, she was transferred to pick up dead birds for West Nile virus testing. That didn't take up all her time, so she began helping in the lab where they tested mosquitoes for disease. One of the other people in the lab had been assigned to learn ArcView® 3.*x* software but got frustrated and gave up. Kate took over and spent time learning the software, eventually making all the maps for the organization.

Kate was working full time because she had been placed on academic suspension from the Computer Science Department at school. It was right after the dotcom crash in the early 2000s so many of her friends were unemployed. She had been skipping school to work on tech projects with them and not keeping up with all her homework. But her switch to geography stood her well.

With a passion for geography to allow people to apply location to many different industries and skill sets to solve myriad problems, Kate has worked in many fields herself, including medical geography, disaster response, land rights, and location-based geography. These days she owns her own company, Cascadia Technical Mentorship, in Winlock, Washington, and credits her love of geography for her work with nonprofits to help marginalized communities make use of open source software and data. Currently, Kate advises a variety of organizations on how to better engage with volunteer open source contributors, as well as tech strategy and general nonprofit management.

Kate considers herself lucky to have had supportive parents who nurtured her and her siblings' interests. Her mom was a science teacher, and they were always performing experiments outside in nature. Her father was an electrical engineer, which propelled her interest in computer science.

Still, school was difficult for her as a child. She was shy, and her family moved many times for her father's job while she was in elementary and middle school. Although she made lifelong friends in high school, she was bored in class much of the time. She dropped out of high school in the middle of her junior year and went to Northern Virginia Community College. Never a consistently stellar student, she often found more interesting things going on outside of school. She and her friends would "dumpster dive" looking for old computer hardware, and they spent many hours programming games and hacking. Her love of nature and the outdoors developed more outside the classroom than inside it.

She took classes at the community college for two years, and then transferred to George Mason University in Fairfax, Virginia. Kate began to do well in school once she switched to geography.

One particularly memorable class was Economics of Emerging Economies. She was one of the few noneconomics or nonfinance majors in the class.

She remembers:

After the midterms were graded, the professor said, "Who is Kathleen Chapman?" I panicked because I was only "Kathleen" when I was in trouble. I raised my hand, and he said, "You got the only A." The class required a lot of writing about economics but you didn't have to do much of the math behind it. I figured my strength in writing may have given me an edge over the econ or finance majors.

With all of today's encouragement of women entering the STEM fields, Kate feels that it's important to keep women in STEM careers once they start. Many women enter STEM fields of study, and then leave midcareer. She says she wishes she had known earlier about some of the frustrations women might face midcareer so she could have developed a better mentorship and support network for herself.

Often, she found she was the only woman in the room or on the conference call. Her husband, who has attended conferences with her, says when people ask him detailed questions, he often must point to her as the expert.

Where she feels most fulfilled is working on diverse teams with a good mix of perspectives, and being part of a community has influenced her greatly during her career. Her most memorable moment came after the 2010 earthquake in Haiti. She was amazed at how the OpenStreetMap (OSM) community jumped in and started mapping to alleviate the chaos. It started in Germany and Japan, with OSM enthusiasts figuring that first responders would need data. Eventually, it led to a group, including Kate, helping the Coast Guard use the data. The collaboration left Kate inspired that people all over the world could come together using simply mapping and GIS to help people in need.

The next major highlight of her career was working at GeoIQ. This was the first time she was involved in a public-facing website that anyone could use. Prior to that, she worked at another geographic data web portal, but it was subscription based so that only those who bought a subscription could make use of her work. She enjoyed working with colleagues to make maps that other people could use online.

Kate attributes her success to tenacity and never giving up. She would spend hours trying to fix systems when everyone else would quit. She continually submitted grant proposals despite numerous rejections. Taking calculated risks has led to many opportunities. She quit her job to work in Indonesia for two months, which eventually turned into four years. Later in her career, one of her most rewarding projects was working with many partners to map the city streets of Jakarta, Indonesia. She and her team brought together hundreds of government staffers to map neighborhood boundaries using OSM. This collaboration allowed the Jakarta government for the first time to distribute street maps of the flooding while it was happening, and not after the fact. This experience led Kate to cofound the Humanitarian OpenStreetMap Team, an offshoot of OSM geared toward disaster preparedness and response, which gave her much of the experience she uses today.

And though Kate says she faced the same obstacles as many women getting ahead in their careers, she also feels fortunate that people backed her up in her endeavors.

I am lucky. My parents paid for most of my schooling. People don't ask me if I have a degree, nor is it needed for what I do. The knowledge from the degree is required though. There have been some jobs I stuck with, which later led to other things of value.

Work environments often reward behavior typically thought of as male, Kate says, and the working population needs to speak up when they see injustices in favor of inclusivity.

Most recently, she took a risk leaving a stable position to start her own company. She is also working on setting up a farm in the Seattle, Washington, area, which is taking most of her free time. "Life is too short to spend all your time working. I'm privileged that I can make that decision," Kate says.

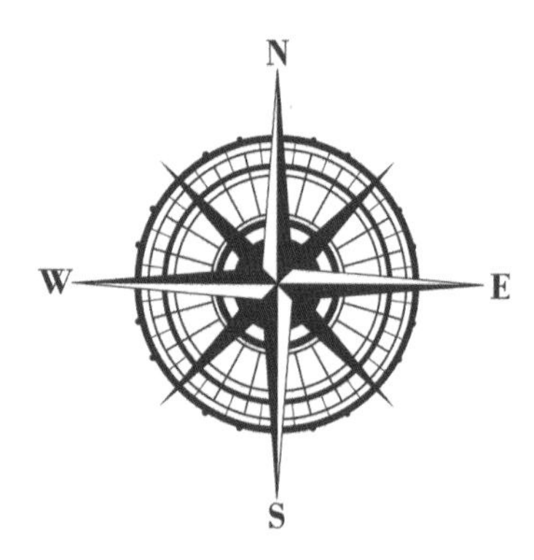

Ultimately, she would like people to better appreciate the significance of crowdsourced geospatial information and to realize that anyone can join, and benefit from, the OSM community. People who use the data or do the research are often looking in from the outside, she says, but OSM is meant to be the map that everyone can edit from the inside.

Kate would also like to see more crowdsourced information being fed into authoritative data sources. This real, on-the-ground information can help augment standard datasets and provide a broader, real-time view, she says. And that's her goal—to make that type of real impact in the future.

Crowdsourced data can allow more eyes on the information so changes can be more readily picked up. This doesn't mean there isn't a place for authoritative data as the official dataset. By working with the two together, we get the best of both worlds. ✸

Wan-Hwa Cheng

Green sea turtles the apple of her eye

SEA TURTLES MAKE WAN-HWA CHENG HAPPY—"THEY ARE AMAZING creatures that are vital to the ocean ecosystem. Without that species, the ecosystem would be significantly and negatively impacted." Wan-Hwa began to learn about sea turtle biology, ecology, and conservation when she was just seven years old. Her father, Dr. I-Jiunn Cheng, is a sea turtle biologist, known as the "Father of Green Turtles" in Taiwan. Throughout her childhood, she visited the nesting beaches around the island with her father's students and listened to them discuss their research.

By the time she was 26 and a graduate student in Florida, Wan-Hwa realized that her father had lots of data with GPS locations, such as satellite tracks showing sea turtle nest locations, but no one had ever used GIS to analyze it. Her father needed help developing maps, and Wan-Hwa decided to test her research ideas on the data by using ArcGIS® software. Having started creating beautiful maps, she's found that she doesn't want to stop.

Currently, Wan-Hwa is a GIS and data analyst, project manager, and consultant for the International Union for Conservation of Nature (IUCN) Species Survival Commission (SSC) Marine Turtle Specialist Group (Taiwan) and a GIS analyst on-site vendor at Apple in Sunnyvale, California, via Apex Systems. In both roles, she uses GIS, programming, statistics, and data science to solve real-world problems, such as understanding the spatial distribution of sea turtles in the nearshore waters of Taiwan, and the interaction between turtle movement behaviors and ocean environment on a three-dimensional scale.

People told her that GIS was difficult to learn, but Wan-Hwa never has been one to shy away from a challenge or hard work. Born in the United States, she moved to Taiwan at age five and returned 21 years later to become a part-time graduate student at the University of Central Florida. Her English was poor at that time, so to improve, she woke up at 5 a.m. every day to study and read. She also worked 35 hours per week as a hostess and cashier at a Chinese restaurant, often writing and designing research ideas on napkins.

Wan-Hwa, *front row right in white blouse*, and her colleagues at Apple in Sunnyvale, California. Wan-Hwa's team works together closely on their projects, and she likes the support she feels in a team environment.

At the end of 2014, as a master's degree student at the University of Central Florida, she was asked by Dr. Katsufumi Sato, from the University of Tokyo, and Dr. Junichi Okuyama, from Kyoto University, to give a presentation to their classes—and after that, she gave the same one to her father's students. She was invited to become one of the authors of "Ocean Says," which is the biggest ocean science Facebook site in Taiwan. *National Geographic* magazine (Taiwan branch) invited her to write an article about how people can reduce the negative impacts of artificial light on sea turtles by using GIS. That experience propelled her to attend her first Esri User Conference in 2014 in San Diego, California. Surrounded by 16,000 like-minded others in this hotbed of GIS enthusiasts, she decided to make GIS her career.

Wan-Hwa wanted to keep learning more to advance sea turtle research. She went to as many regional and international symposiums as she could. Attending symposiums has provided her with opportunities to network with sea turtle biologists from all over the world, learn new ideas, answer questions in oral presentations, and help organizations from other countries with GIS. In April 2015, she became a student evaluator and session co-chair at the International Sea Turtle Symposium in Dalaman, Mugla, in Turkey.

When she first started to attend symposiums, she lacked confidence and didn't know how to interact with other students. She remembers attending a regional meeting and standing alone during the social event, waiting for other people to come talk to her. Frustrated, she called her friend, who gave her inspiring advice. Her friend reminded Wan-Hwa that social interaction was different in the United States, but that she needed to overcome the obstacles to approaching others. She went back to the conference room, listened to some presentations, and, gathering her courage, started to discuss her research with other students. Once Wan-Hwa decided to be brave and not make excuses, the ice was broken. Now she can attend conferences and talk to others easily.

In early 2016, she transferred to the master of advanced study in GIS (MAS-GIS) program at Arizona State University, which allowed her to fully develop her GIS skills. All the while, she continued sea turtle research with GIS, and presented her projects at the Esri Ocean GIS Forum in Redlands, California, in 2017.

Wan-Hwa gives a presentation at the International Sea Turtle Symposium in Las Vegas in 2017.

Of course, she encountered obstacles along the way, including one of her own making—often overworking herself to find the answer to a problem.

Some people laughed at me and tried to discourage me. Even some friends did not understand my dedication to academics and research, or underestimated me, as in "helping" me do my poster presentation in the school conference without asking [my] permission. This bothered me when I was younger, but as I matured, I decided to celebrate how I am different by continuing to do what I love and enjoy.

After graduation in 2017, master's degree in hand, Wan-Hwa began her work as a GIS and data analyst for the IUCN SSC Marine Turtle Specialist Group. In the beginning, she helped map the sea turtle satellite track using two-dimensional data that tracked longitude and latitude. Using spatial analysis, she began to train the graduate students from the Institute of Marine Biology, National Taiwan Ocean University, on their GIS skills and helped with spatial analysis for the Taiwanese government. Currently, she is working on understanding the interaction between the three-dimensional ocean environment and sea turtle movement behaviors, focusing on the spatial distribution of the sea turtle population in one main foraging ground in Taiwan.

As Dr. Dawn Wright [an oceanographer and chief scientist of Esri] points out, the ocean is a three-dimensional environment. If we use only two-dimensional data, we probably cannot provide the best suggestions for strategizing sea turtle conservation. Therefore, recently we began to use three-dimensional satellite tracking data, which includes longitude, latitude, and dive depth. This is important since sea turtles do not [swim only] on the sea surface, and the plastics and trash [are not only on the surface] either.

Wan-Hwa's IUCN team in Taiwan uses science to help the Taiwanese government protect sea turtles and increase sea turtle populations. The team has been doing sea turtle research in Taiwan for more than 20 years on both the land and in the ocean. On land, they help the Taiwanese government build marine turtle rescue centers for the diseased and damaged turtles that people find. Every year they conduct a beach survey to count the number of nests and measure the size of nesting female turtles. Wan-Hwa has also participated in investigations to determine how certain beach environments affect turtle eggs and turtle health.

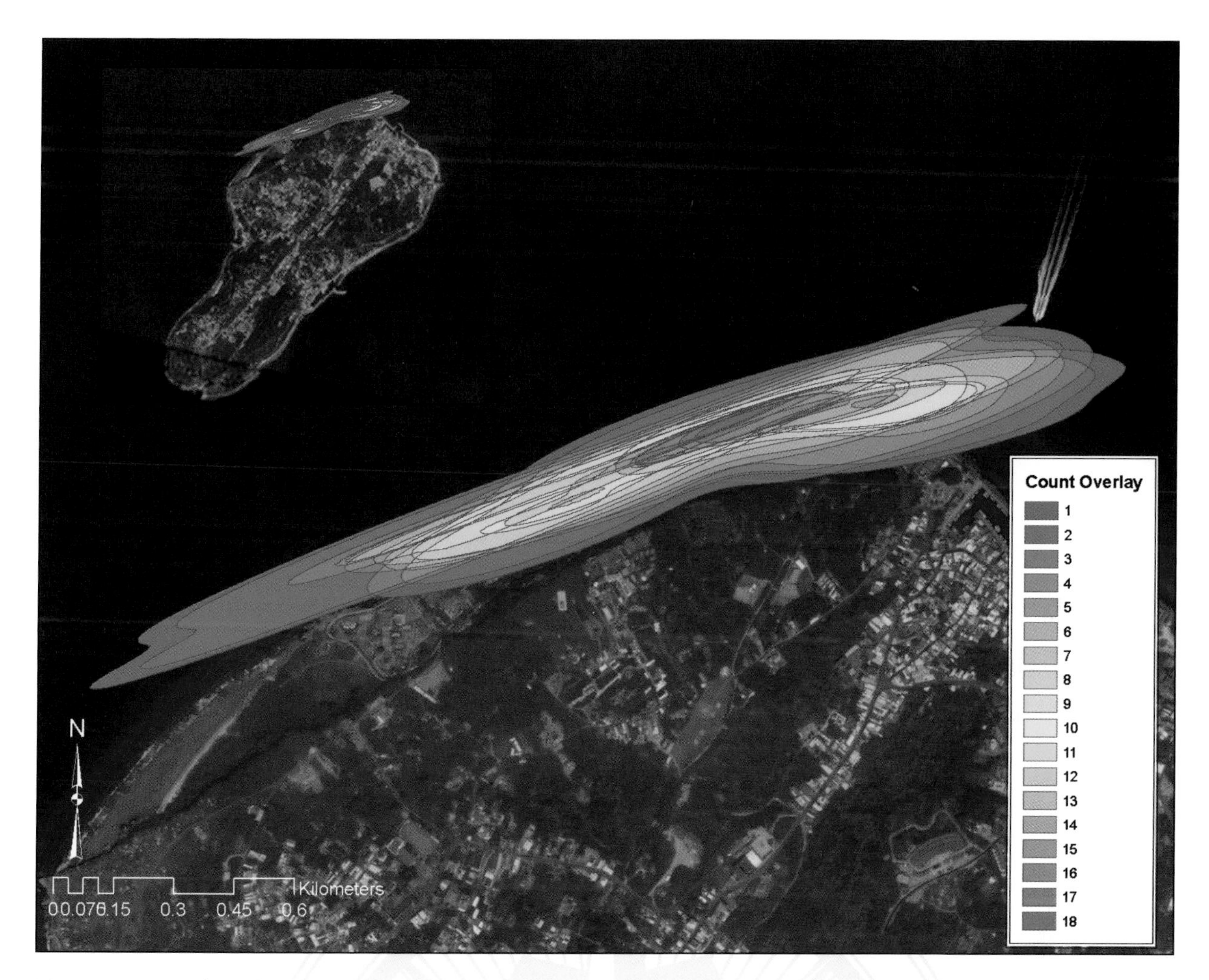

Wan-Hwa's map shows her analysis of sea turtle home range. She presented this map at the International Sea Turtle Symposium in Las Vegas in 2017.

Photo taken during state and federally permitted research activities.

Wan-Hwa with a giant leatherback turtle at night after the team had finished taking all the measurements and was waiting for the nesting female to return to the ocean. Leatherbacks, the largest sea turtle species, can measure up to nine feet and weigh over a thousand pounds.

In the ocean, the team analyzes satellite tracking and in-water surveys to understand how the ocean environment influences green turtle swimming behaviors. They estimate population size and home range around the island of Taiwan, often collaborating with NOAA and sea turtle biologists in Japan.

All of Wan-Hwa's projects require spatial analysis, data science, sea turtle biology, oceanography, statistics, and programming skills—all STEM skills, which traditionally females worldwide have not been encouraged to learn. Having a father with the celebrity status of a rock star in his country for his scientific prowess in sea turtle study, Wan-Hwa managed to sidestep most discouragement. In fact, when working and traveling globally, she rarely makes it known that she is this famous man's daughter.

Many teachers, students, friends, and biologists encouraged me to pursue my passion. Yet, growing up in Taiwan, I did encounter people who think men are better than women and [that] women should not receive higher academic degrees but be more marriage-minded.

For her current success, she credits the support of her family, especially her parents and grandparents, and their belief that men and women are equal. Her father is still a major influence and supporter of her work. Wan-Hwa uses the long-term datasets from his laboratory to test her research ideas. "I love discussing my research with my father," she says, "because he is always honest and helpful."

She points to other career mentors, the professors and students from the Department of Biology, University of Central Florida, as key to her learning to develop and design research projects independently: "to be able to ask good research questions, to think about how to use scientific methods to solve real-world problems, to convince and explain research ideas, and to get information from a large number of publications in a short period of time." She also credits the professors and lecturers from Arizona State for showing her "how ArcGIS and programming languages can help people to solve their problems with real examples."

Wan-Hwa's love of research to learn more about important ecosystems using technology's tools is ongoing. It rewards her daily with the same delight that came at her first sight of a sea turtle, and she feels moved to share it.

Wan-Hwa's map, *left*, tracks a loggerhead sea turtle, #107366, that was released at Dong'ao, Yilan, Taiwan, on February 22, 2014, and is still being tracked. She used a map from the past, *right*, as the basemap to show that sea turtles already existed in the era of dinosaurs.

I tell young women that if they dream of helping the environment or saving endangered species, they should enter the STEM fields. It's my hope that in the future, people will use what they learn from my work to make the world a better place. ✸

yaichatchai/Shutterstock

Sylvia A. Earle

Her Deepness works to save the ocean

Photo courtesy of Kip Edwards and Mission Blue.

Sylvia diving at Cabo Pulmo, located in the Gulf of California Hope Spot. The community was once sustained by fishing but now is collaborating with biologists, conservationists, government staff, and divers from around the world to create a no-take marine reserve. This effort gives Sylvia great hope for the sea.

WHEN SYLVIA A. EARLE FIRST PLUNGED INTO THE DEEP, THE OCEAN WAS Wonderland and she was Alice. The world-renowned oceanographer feels much the same now, still scuba diving at age 83, her wonderment intact. The world's oceans are not the same, however, coming under catastrophic threat during her lifetime, so now her mission is to save the oceans. Human survival depends on it.

By now, Sylvia the explorer has led more than 100 expeditions and spent 7,000 hours under the sea, earning the appellation Her Deepness from the *New York Times* and the *New Yorker*. In 1979, she walked solo on the ocean floor, a quarter of a mile deep, the first person to do so. As ocean pioneer, she has seen it all, but even she didn't see this crisis coming.

Like everyone else, Sylvia assumed the ocean was too big to fail, yet in the last few decades, from overfishing, global warming, plastic pollution, and climate change, half the planet's coral reefs have died, and more than 90 percent of big-fish populations such as bluefin tuna, cod, and shark have disappeared. Species thriving since the age of the dinosaurs, such as sea turtles that require multiple decades and thousands of migration miles to mature, mate, and feed, are in sharp decline, even on the brink of extinction.

With a lifetime of research and experience—as marine biologist and botanist, National Geographic Society explorer-in-residence, and former chief scientist of NOAA—Sylvia knows what she's

Even as a young adult entering the occupation, I never imagined–no one did–that the oceans, 75 percent of the earth's surface and why we call our home the Blue Planet, were vulnerable.

talking about. In a 2009 Ted Talk where she accepted a $100,000 Ted Prize, Sylvia says, "Fifty years ago, when I began exploring the ocean, no one—not Jacques Perrin, not Jacques Cousteau or Rachel Carson—imagined that we could do anything to harm the ocean by what we put into it or by what we took out of it. It seemed, at that time, to be a Sea of Eden, but now we know, and now we are facing Paradise Lost."

The ocean is the earth's largest ecosystem—driving weather, regulating temperature, shaping planetary chemistry. But so much of this living system is still largely unknown. According to NOAA, "More than 80 percent of this vast, underwater realm remains unmapped, unobserved, and unexplored" (https://oceanservice.noaa.gov/facts/exploration.html, August 30, 2018).

With so much of the ocean still so full of mystery, Sylvia might have preferred to keep her focus solely on discovering more of its underwater secrets. Yet she is intent on helping others see it through her eyes.

Dr. Sylvia A. Earle, *Time* magazine's first Hero for the Planet (in 1998) and a Library of Congress "Living Legend," takes her mission of advocacy on the road now, 300 days of the year, so that we all can see what it's like to dance with an octopus, to stare down a fish looking back at you, to swim beside a 20-ton, 40-foot whale shark. All of this to inspire us to act before it's too late. Humanity has put a lot of resources into looking toward the sky since the space age, but not enough into looking down below, to understand the sea around us.

We have creatures that are just so exotic. Earth is where the strange and wonderful creatures exist! You don't have to use your imagination, you can just dive into the sea.

We have learned enough, though, to realize that we are changing the chemistry of the sea. Just in the last 10 years, "we are seeing broad low-oxygen levels in the ocean," she told *Huffington Post* in 2017. "That makes it hard for the fish to breathe, and so much more." Using her Ted Prize award, Sylvia founded the nonprofit Mission Blue in 2009 to educate us about the importance of the ocean. Its Hope Spot Initiative supports establishing marine sanctuaries all over the globe—Hope Spots—to help heal, protect, and sustain the ocean.

Mission Blue (Mission-Blue.org.) works with local champions of marine conservation around the world. As president and chairman of Mission Blue, Sylvia sounds the alarm that only 3 percent of the ocean is fully protected.

Another 4 percent has some protection, but overall, much more needs to be safeguarded to maintain the integrity of vital ocean systems.

Mission Blue's near-term goal is to reach the United Nations' Sustainable Development Goal of protecting at least 10 percent of the ocean by 2020. To get there, it's partnered with more than 200 ocean conservation groups and led or joined Hope Spot expeditions to many locales. Plans are in progress to further "embrace GIS technology and story mapping as a powerful narrative device to further raise awareness of Hope Spots and the stories that emanate from them," she writes in the 2017 *Hope Spots Impact Report*.

Sylvia and the Mission Blue team declared the Balearic Islands as the first Mediterranean Hope Spot. The Balearic Islands, including Mallorca shown here, host a good representation of the habitats along the Mediterranean, as well as high biodiversity.

To facilitate marine research, she helped develop a small submersible called *Deep Rover* that she describes as "so simple even a scientist can operate it." This transportable craft surrounds the operator with an atmospheric shell much like an astronaut in a space suit. The vehicle, which provides up to 100 hours of life support, doesn't require operators to undergo a decompression period and is ideal for conducting research in the sanctuaries. The cofounder of Deep Ocean Exploration & Research Inc., Sylvia herself has used about 30 kinds of submarines and collaborated in creating many such life support systems.

On a recent dive into a Hope Spot, Sylvia emerged from the depths to report what she'd seen. She rarely speaks of looking into the eyes of the ocean's underwater inhabitants without pointing out that they are looking back into hers. She has cruised the undersea world for more hours than any other human being—7,000 hours—and become an award-winning photographer for capturing its intensity.

Sylvia first fell in love with the ocean when a wave knocked her off her feet during a family vacation at the Jersey Shore. Excited by the waves, she rushed in for more and hasn't stopped since. When she was 12 and her family moved from New Jersey farmland to Clearwater, Florida, the Gulf of Mexico's clear, warm water became her backyard.

Graduating in 1953 from being simply a swimming teenager to one who snorkeled transformed her "irreversibly into a sea creature who henceforth would spend part of the time above water," she writes for young readers in *Dive! My Adventures in the Deep Frontier*, one of her many books. "It's when I first got to know fish swimming in something other than lemon slices and butter," she says. Like her scientist contemporaries, primatologists Jane Goodall and Dian Fossey, Sylvia was determined to study living beings in the habitat where they lived. She started with plants, earning her BS degree from Florida State University in 1955 and her MS and PhD degrees, respectively, in 1956 and 1966 from Duke University. "It took me 10 years to get my PhD because I got married and had kids along the way," she says. "It was always a juggling match." But the call to the sea was irresistible.

The Hope Spot at Cashes Ledge is home to the largest cold-water kelp forest on the Atlantic Seaboard. Courtesy of Kip Edwards and Mission Blue.

I was just enchanted with the idea of getting into a little ball, with a tiny little window, and descending a half mile below the surface, where it was really dark all the time. Except that the way that [diver and] author William Beebe described this galaxy of life was luminous: the flash, sparkle, that glow of bioluminescence—that firefly kind of light that is common in the deep sea. I thought, I want to do that!

Sylvia became one of the first divers in the United States to try scuba gear, then called the Aqua Lung, in the early 1960s as a graduate student at Duke University, on her way to getting a PhD.

The summer of 1964 changed her life. Only four years earlier, Jacques Piccard and Donald Walsh had descended to the deepest part of the sea, 6.78 miles down. From their submersible, they discovered never-before-seen forms of life in the Mariana Trench. Revolutionary change would follow, and Sylvia would take her place as a key player right in the middle of it.

In 1964, when one of Sylvia's colleagues at Duke had to drop out of a planned expedition to the Indian Ocean for six weeks, grad student Sylvia, also a young wife and mother at the time, was invited in his place. They would be diving where no one had gone before, yet she herself had never been out of the country, or even west of the Mississippi River. Shortly before departure to Mombasa, Kenya, she got a call from the chief scientist of the expedition, Ed Chin. He said, "This may not be a problem, but you should know, you're going to be the only woman on board, and there are 70 men." It proved no problem for Sylvia.

Interviewed in advance by the local press, the scientists poured their hearts out about the fascinating work ahead, and the limitations of exploring the average 2.5-mile depth mostly from the deck of a ship ("with these pathetic little tools to try to sample this huge expanse of living blue," she recalls). Yet—in a sign of the times—the next day the newspaper's headline read, "Sylvia sails away with 70 men," with the subhead, "But she expects no problems."

The expedition was a success and set the stage for the next major step in Sylvia's calling. In 1970, at age 35, she was chosen to lead an all-women team of aquanauts living for two weeks in an underwater habitat, called Tektite II, off the Virgin Islands. She'd seen the call for applications on a bulletin board. "No one expected women to apply, but we did," says Sylvia. Again, the initial newspaper stories asked such things as, How will women fare, sharing such close living quarters? Will they wear lipstick? The newspaper headline as they started their mission read, "Five Gals Face Plunge with One Hair Dryer."

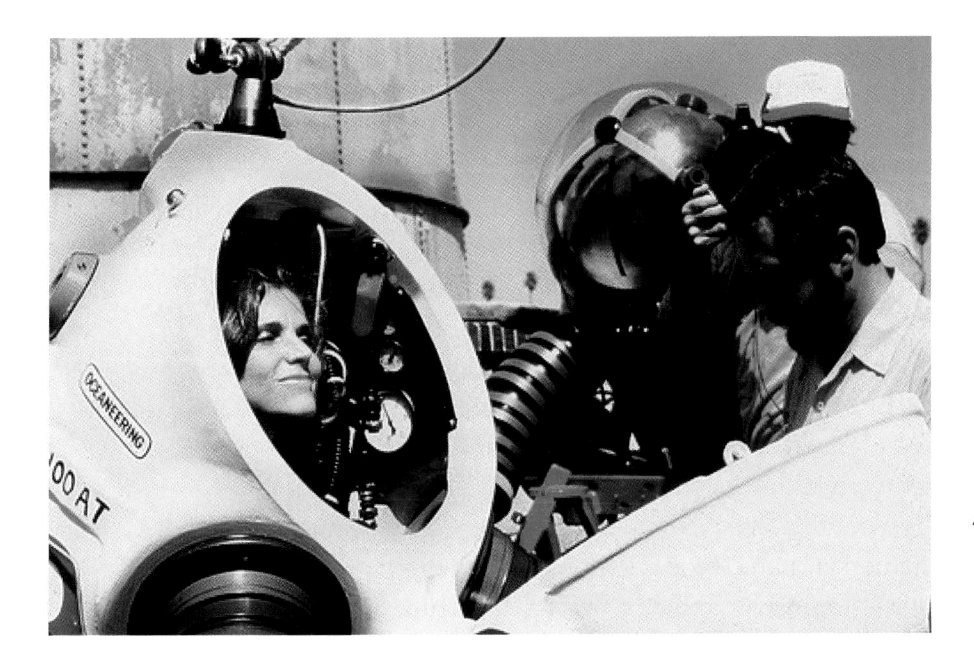

Off the coast of Oahu, Hawaii, Sylvia made her human depth record-breaking solo dive in JIM, "one of my favorite bathing suits," she quips. Courtesy of NOAA.

Yet again, Sylvia (and her four teammates) performed admirably. The official report of their success at the undersea laboratory Tektite II says, "Their fine performance paved the way for the routine inclusion of female aquanauts in future undersea missions and female astronauts in space missions." That first for women put her on the cover of *Time* magazine.

What followed ultimately led to her work as a *National Geographic* explorer and author, her stint as the first female chief scientist of NOAA, and her entrepreneurship in starting three companies to build submersibles and two nonprofits. Astronauts were putting their footprints on the moon in 1969. In 1979, Sylvia climbed into a special diving suit called a JIM suit and put her footprints on the ocean floor. She walked freely six miles offshore and 1,250 feet down (381 meters)—a record that still stands.

In this suit that allowed her freedom of movement away from any kind of enclosed submarine, she was like an astronaut on a spacewalk, except that her realm was inner space rather than outer.

With her wittiness and matter-of-fact demeanor, Sylvia broke through barriers as if she hadn't seen them. Sometimes she avoided obstacles by removing herself from their path, as when she resigned as NOAA's chief scientist in 1992 after only a two-year stint. Quickly, she'd realized that she would not be able to use her voice for change as effectively as she might outside a government (or corporate) institution.

Her message, which calls like a shell pressed to the ear: beauty abounds in the ocean as does mystery. Moreover, our oceans can rebound and recover, if we open our eyes and help restore what we've so blindly depleted.

Sylvia's walkabout on the ocean floor four decades ago gave her the moniker Her Deepness, and it reflects her deep and abiding love of the Deep. In the award-winning documentary film *Mission Blue*, which begins with her pointing out devastation from the disastrous Gulf of Mexico oil spill in 2010 more than 60 miles away from its originating leak, filmmaker James Cameron refers to her as the "Joan of Arc of the ocean."

Sylvia uses GIS to help others see what imperils our vital ecosystem.

Where does GIS come into all of this? Better to ask where GIS doesn't play a role in understanding the ocean.

Everything about marine ecosystems is geospatial in nature, Sylvia explains. Spatial, temporal, and functional relationships among creatures, environments, and human perturbations must be recorded and integrated to arrive at meaningful solutions. In the last two decades, Sylvia says, improvements in GIS tools have provided a "quantum leap in our ability to understand and manage marine ecosystems." For Sylvia, foremost among these new tools is three-dimensional GIS data modeling, particularly of ecological marine units (EMUs).

Sylvia has always been a fan of what technology can do—and help undo. She led a five-year exploration called Sustainable Seas Expeditions (SSE) in 2002, using GIS to study US national marine sanctuaries. Mission Blue uses the technology to identify new Hope Spots and help maintain the official marine sanctuaries that countries around the world have committed their support to. GIS gives SSE an improved basis for planning and conducting future expeditions as well.

In combination with daily advances in remote-sensing imagery and the storehouse of data accumulated by satellites over the last 60 years, GIS has helped put the writing on the wall in terms of the perils the ocean faces unless we act to save it.

"Now suddenly so much I can see, as never before, in context," Sylvia says. "It's that kind of aha breakthrough that these new integrating data systems allow." Add to that the power of the internet and social media to rally support on a global scale.

We have the power to change the world–the way we look at the world, the way the world goes forward henceforth–if we just use the technologies and join together with our minds and our hearts and our commitment to make a difference. ✸

Shoreh Elhami

A life of service and volunteerism to the greater community

FOR AS LONG AS SHOREH ELHAMI REMEMBERS, SHE WAS INTERESTED IN both science and math. She had several great teachers in middle school and high school who encouraged her. At home in Tehran, Iran, her mom and dad instilled the importance of a lifelong desire to explore, learn, and experiment, in both her brother and herself. She feels that her proficiency in math and science played a key role in her becoming a successful woman and immigrant in the STEM fields and fueled her desire to pay it forward to the next generation of female leaders in STEM.

When she was very young, she wanted to become a mechanical engineer. She loved spending time with her dad on his car, changing the oil or a flat tire. She felt deeply, even at a very young age, that there wasn't anything that she couldn't do.

In Iran, students had to choose a major before starting the 10th grade—math/science, biology/chemistry, or literature. Courses taken in grades 10 through 12 would then focus on that major. Shoreh majored in math/science and, by starting early and skipping a grade, graduated from high school at age 16.

Shoreh then started studying architectural engineering in 1977 at the National University of Iran in Tehran. Because of the Iranian Revolution, all universities closed in 1980–1981 for almost a year.

Before the universities closed in Iran, Shoreh, *top row, far left*, during her sophomore year in 1979 on a field trip.

It was difficult to want to be in school to learn but not be able to do so. When they reopened the university, students were required to take mandatory and mostly irrelevant courses, which made it unbearable for those like Shoreh who wanted to learn new things, but she went back and finished her degree. After graduation, she worked as a consultant and for a brief period worked in the southern region of Iran, which was torn by the Iran–Iraq War. She and her crew would go to bombed-out areas to record the details of devastation, and then generate revitalization plans for rebuilding the affected areas. She has vivid memories of being called "Mr. Engineer" by the locals, despite clearly looking like a woman. In some of those remote areas, seeing a woman in charge or in a position of power was simply unfathomable to the local people.

Shoreh, *seated third from right,* in class in 1980, after it was made mandatory to wear the hijab.

In 1985, Shoreh and her family left the country shortly after she graduated from college, and she came to the United States to go to graduate school, this time studying city and regional planning. Her first few years in the United States were not easy. Her family didn't know anyone and didn't have any family members close by. She flew into Gainesville, Florida, with an infant in her arms, and even though she and her husband had studied English at school, they realized their knowledge of the language was inadequate. She stayed home with her daughter for two years, and then started graduate school in 1987. The hardships she faced weren't all that different from those that all young women encounter when they want to go back to school or the workplace but also want to be with their children.

After attending graduate school at The Ohio State University, Shoreh started working at the Delaware County Regional Planning Commission in Delaware County, Ohio, where she was charged with building a GIS to enable the agency's planning functions. After that job, she worked for the Delaware County auditor's office as GIS lead. Currently, she works as the data, analytics, and GIS lead for the City of Columbus Department of Technology in Ohio.

Shoreh's work, though, extends far beyond Ohio. An active member of the Urban and Regional Information Systems Association (URISA) for the last 29 years, Shoreh is the founder of URISA's GISCorps, whose mission is to coordinate short-term, volunteer-based GIS services to underprivileged communities. She is also a former member of the URISA Board of Directors. Along with hundreds of volunteers, she and GISCorps' Core Committee contribute their time and efforts to various missions, including supporting humanitarian relief, enhancing environmental analysis, and encouraging economic development. The GISCorps recently launched its 250th project. To date, over 2,000 volunteers have contributed thousands of hours to communities in need around the world, volunteering their technical expertise for disaster response, to assist in community building, and to support sustainability.

In the mid-1990s, Shoreh's county, Delaware, was among the first in the United States to publish an interactive map providing detailed cadastral data to the public and to openly publish current and accurate GIS data. Columbus was one of the first cities in the United States to experiment with ArcGIS® Hub from Esri® for an infant mortality initiative. In Columbus, GIS is used in almost every department, including public safety (911), public service, utilities, planning, economic development, neighborhoods (311), recreation and parks, housing, and building/zoning. Besides these efforts,

Shoreh also taught a graduate GIS course for 10 years at The Ohio State University and has coauthored and/or contributed to several books on the US Census, cadastral mapping, and the role of GIS in disaster response.

Shoreh, in volunteer work for the GISCorps, teaches GIS to students in Kabul, Afghanistan, in 2005.

Shoreh says that women have made progress in the past 30 years in the STEM fields, but there's still a long way to go. Her focus is on the root causes of the gender gap and on eliminating inequalities in a systematic way, which starts both in schools and in legislation.

We need to educate women about gender equality and their rights, and elect lawmakers who fight for them.

Shoreh wants to further the inroads of girls in the STEM fields and nurture girls and young women in these fields by mentoring them and sharing their success stories with others.

I also mentor high school girls and will be starting with two new mentees, who are both attending a local high school that's in dire need of support.

One of Shoreh's most gratifying rewards is witnessing the girls and young women that she has mentored now mentoring other women. Her love and enjoyment of math and science has proven an effective way for her to attain equality in pay and status and has empowered her to have a successful and fulfilling career in the STEM fields. ✷

Shoreh, *center*, with students and colleagues in Kabul.

Karen E. Firehock

Connecting people to green infrastructure

Karen visits a stream in rural England as she collects images of water management and stream health to use in her teaching.

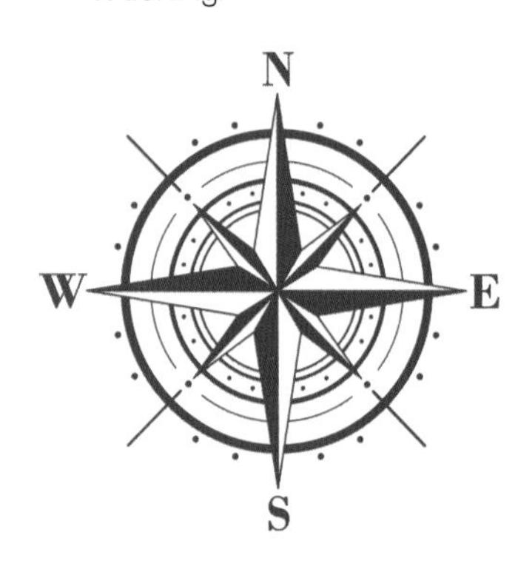

SURROUNDED BY NATURE—TREES, STREAMS, FLOWERS, ROCK OUTCROPS, any place in nature—is where Karen Firehock wants to be. Spending most of her childhood playing or working outdoors, it wasn't until she realized that she could "work in the field" that she became focused on her career path. Karen understood her passion early on and built her career around it. Hers is a story of passion meeting personality because Karen has the natural instincts to create solutions, build networks of support, and ultimately connect people with nature.

Karen is an environmental planner who runs a mapping and environmental company, overseeing green infrastructure planning and research. She is also a lecturer in green infrastructure planning and environmental ordinance development at the University of Virginia's School of Architecture in the Department of Urban and Environmental Planning. Karen works to keep green infrastructure—natural resources such as forests, waterways and bays, soils, wildlife areas, wetlands, dunes, historic landscapes, and parks—at the forefront in local planning. That way, green resources are considered along with planning for population growth, road plans, and traditional infrastructure costs, the normal purview of local governments. She also works as a writer, researcher, and facilitator for environmental negotiations, all in behalf of green infrastructure.

Karen's earliest memories of her fondness for nature are of hiking in the woods with her father. Every Sunday Karen's dad took her and her brothers outside to explore a forest, hike a mountain, or just run through Rock Creek Park, which spreads across Washington, DC. Her grandfather taught her to sail, swim, and snorkel, and her grandparents took her to the Everglades in Florida and other special places. When Karen was eight or nine, her parents bought a 20-acre farm on the Eastern Shore of Maryland in the early 1970s because her father needed a break from his high-level job as an analyst at the Arms Control and Disarmament Agency, where he worked on denuclearization of the United States and the Soviet Union. The family spent most weekends at the farm, along with time in the summer and over the holidays. Each of the four children helped with the farm, mostly growing vegetables, but they were each given their own plot of land to plant and harvest.

Karen's mother, a painter who loved to paint scenes of nature, taught the children to be little entrepreneurs. Karen and her brothers set up a red wagon to become a kids' farm stand in front of Riggs Bank in DC, where they sold produce. When they were not tending the farm, the kids were roaming and exploring the surrounding woods and wetlands.

In Karen's high school, students who were good in math took physics and stayed in the classroom. Since she had a stronger aptitude for science than math, she moved to the earth science class instead.

When I realized that people had jobs to go outside–field science–I realized I could combine my love of the outdoors with my vocation and decided to focus my career on environmental science.

Karen worked her way through a private high school on a *Reader's Digest* scholarship.

Karen went to Allegheny College, a private liberal arts college in Pennsylvania. In her second semester, her parents stopped helping her with tuition, forcing Karen, then 19, to have her grades frozen, return home, and emancipate herself to be able to qualify for tuition assistance. Emancipation required that Karen, a minor, live on her own. Suddenly, she was completely responsible for herself.

Because community colleges did not ask for transcripts, Karen attended one for several semesters while she worked to pay off her previous tuition bill. When she got her grades and transcript back, she went to the University of Maryland, studying geology and then natural resource management. She attended college part time so she could work, and it took her eight years to finish. She worked for the Izaak Walton League of America, a conservation organization, as a receptionist. Additionally, she was asked to answer bushels of letters from children across the country. The letters were written in response to a conservation project called Save Our Streams (SOS) that used a Winnebago to travel the United States and teach how to test local streams using simple chemical kits and a "bug test" that involved looking for three aquatic organisms—stone flies, mayflies, and caddisflies—indicating the water was clean. Kids wrote back saying, "Me and my friends staked out the post office and asked everyone who came there to help us clean up our creek. We held three creek cleanups and formed a 'save the creek' club. We are 8." As Karen responded to each letter, she was inspired to develop more activity kits for the children who showed interest, and her project was sponsored by a partnership between the Izaak Walton League and the CNN show *Captain America*. Soon she was giving the kids Earth power rings and developing even more materials. She also created an adult version of the stream-saving kit.

The Ohio Department of Natural Resources (DNR) was intrigued by the Izaak Walton League's success using volunteer stream testers. The Ohio DNR, which was required by the state government to monitor the state's scenic rivers, needed help expanding its monitoring network, and the two organizations agreed to collaborate. Karen and the DNR created a more scientific protocol using macroinvertebrates to measure stream health and wrote a grant proposal to fund participation that paid for new bug ID cards and expanded testing kits. The Ohio DNR took Karen's approach of using volunteers to monitor and measure the health of scenic state rivers. Its biologists trained volunteers in each of the state's scenic watersheds. The field tests went well, and the volunteers showed they could perform well.

Seeing a government agency use the data, I realized we could augment every state program with volunteer monitors.

Karen created a scientific protocol that the US Environmental Protection Agency (EPA) adopted and a chapter for the method in the EPA's national guide *Volunteer Stream Monitoring: A Methods Manual*. She piloted statewide volunteer monitoring networks in Virginia, West Virginia, and Tennessee, and then developed new protocols for tidal and muddy rivers in Louisiana and other coastal states. She also developed a similar protocol for wetland monitoring. The data that the volunteer monitors acquired was used to highlight streams for restoration, comply with state monitoring requirements, and add streams to impaired lists for funding and cleanup, as well as to find and stop multiple pollution violators. In one community where she taught, a mother had a child who was sick and nobody knew why. Other children in the community had also been ill. "No one could understand why the children were becoming ill. The sickest child was taken to Johns Hopkins for testing, and they found that they had illnesses that can happen when exposed to raw sewage," Karen says.

Our stream monitoring found the sewage treatment plant had failed and had been sickening the children for many months. The children played in this creek along their school's fields. The violation was then addressed, and the children recovered.

As the director of SOS in Ohio, Karen developed a stream restoration program called Stream Doctor and led habitat restoration projects with a national manual, a video, and training workshops. A growing number of citizen stewards of all ages were getting involved in preserving the health of waterways, which was making a difference. Karen's work was recognized; she even was thanked at the White House and spent time in the Rose Garden talking to President George H. W. Bush about streams.

"Over time I realized that I was always monitoring pollution, restoring a damaged habitat, but was I preventing the landscape-scale damage that caused it? I realized that the large paving for malls, highways, et cetera and general sprawl were damaging the rivers and wetlands faster than I could fix them."

To address the heart of the problem, Karen decided she needed to go back to school and become an environmental planner.

In graduate school at the University of Virginia, Karen learned about other landscape designers such as Ian McHarg and Frederick Law Olmsted. She learned that GIS gave her a digital analysis tool that could do far more than she was able to do on her own drawing on topographic maps. Recognizing that land planning is fraught with conflict, she trained as a mediator working for the university's Institute for Environmental Negotiation. After graduating, Karen worked for the institute on multiparty dispute resolution, including settlement agreements for recording the transport of nuclear waste across the United States, ATV rules for national forests, and community-based watershed planning for several communities.

Karen continued to learn as she got involved with other cultures. The Center for Health at the University of Virginia wanted to develop a joint research agenda with the University of Venda in South Africa, and so Karen visited the country as part of a team involved with global water issues. In South Africa, Karen led the workshop to develop the research agenda and engage students, faculty, and community leaders from both universities. Then the team toured five villages to learn about their needs. Karen saw places where water ran only one or two days a month. She learned that people were dying because of a lack of access to clean water, especially those with HIV-AIDS because the contaminated water caused intestinal illness that prevented them from absorbing their medicines. This was a level of poverty that Karen had seen before in her work with communities in Appalachia in Virginia and West Virginia. Access to clean water is a global problem—not just a Third World issue, she realized. But along with the struggles she witnessed in South Africa, Karen also saw a profound level of spirit and caring, especially among the women who volunteered to work with her and help feed and care for the villagers.

A formal partnership was created between the two universities and the villages in the Limpopo region of South Africa, with a project called Water and Health in Limpopo (WHIL). Once the villages agreed to collaborate, Karen packed up her laptop, flip charts, colored sticky notes, markers, and name tags, all tools that she had used successfully in the past when coordinating collaborations. The first meeting was held under the sacred meeting tree, where all important discussions are held.

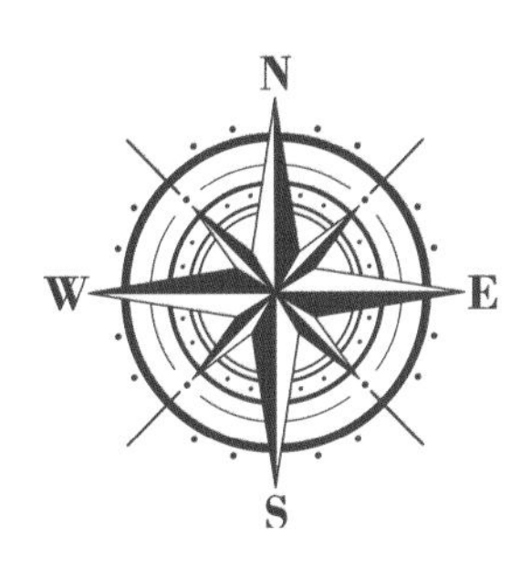

I realized I was not going to put big pink Post-it® Notes on the sacred tree–nor was I going to show up and run anything. The emissary of the village's chief would run the meeting. I would speak when and if called on. I had to fit into their world as a guest–I had to do a lot of listening first.

Another challenge the team faced was the lack of data about the community and its water needs. When apartheid fell and white engineers and staff were pushed out, they destroyed their data about the community. To determine water issues and supply needs, the WHIL team not only had to perform a census but had to remap the villages' streets, wells, and so on. A team of doctors, scientists, and engineers did the analysis, while Karen and a nursing faculty member led the field crews of villagers and college students. She saw firsthand the struggles people had getting clean water. Knowing that they had mostly been left on their own to supply water, the villagers devised a system of rubber hoses to bring water down from the mountain springs, but the water became contaminated along the way. Research by students showed that the water stored in homes also became contaminated, even if it arrived clean, since it sat in open containers and there was no way to store it safely.

Their efforts resulted in having water plumbed into the villages more effectively (more storage and supply lines) as well as perfecting a ceramic filter technology called MadiDrop for cleaning water stored in the home. The two universities collaborated to set up a factory in a nearby village where a women's pottery collaborative was already operating. Local potters were hired to make the ceramic filter technology (http://www.puremadi.org/).

Working as part of a team, Karen came to understand that "my world view was only mine—others had no context for my ideas."

I had to become part of their world first. I had to listen a lot, and then think and then respond. My job was not always to solve the problem—nor could I. My job was to be honest and thoughtful, and then make a difference [when and where] I could.

Karen wrote several guides about environmental collaboration and edited a book about it. She also taught at the University of Virginia and developed graduate courses in watershed planning, codes and ordinances, and site design. Then she read a book called *Green Infrastructure: Linking Landscapes and Communities*, by Mark A. Benedict, Edward T. McMahon, and the Conservation Fund (Island Press, 2006), and adopted it for her classes. This book had a huge effect on Karen: it changed her world view.

Karen demonstrates how to plant a mango tree at an elementary school in commemoration of Nelson Mandela Day. The school is in a rural village in the tribal Venda region of northern South Africa where Karen and her students developed curricula to teach children about watersheds, clean water, and hygiene as part of the WHIL Province Project. The children then participated with Karen to plant more mango trees around the school to provide shade and sustenance.

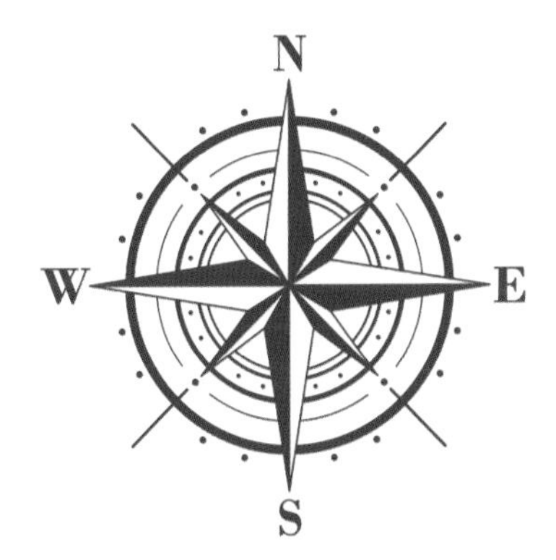

I realized if I could teach communities to see nature as "green infrastructure," to clean the air and water, provide recreation, and support agriculture, then I could get localities to plan to conserve and reconnect natural landscapes as part of everyday planning. I was going to teach this class and save the country. Then I realized I would be long dead before I came close, and so I started an NGO to enlist more partners called the Green Infrastructure Center Inc., which just celebrated 10 years.

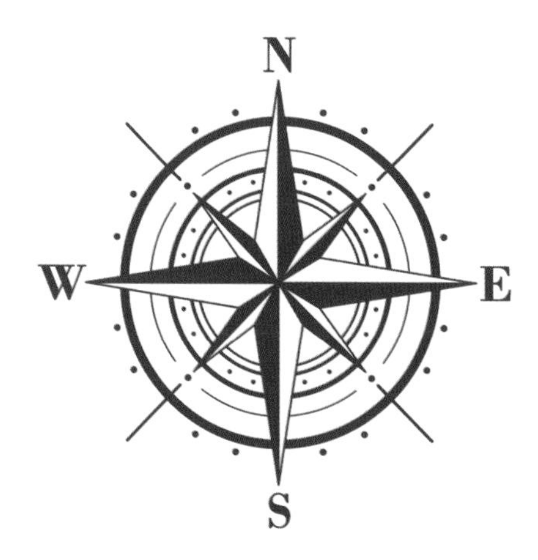

Karen inspects a storm water pond in Charlottesville, Virginia. This pond is designed with plantings to cleanse storm water and provide bird and pollinator habitat.

Before she started the Green Infrastructure Center (GIC), Karen had begun work on her PhD. Now she had to make a choice: finish earning her PhD in collaborative decision-making or go back to her environmental field roots. She tried to run the GIC, working 50 hours a week, while spending every evening and weekend studying for her PhD. It soon proved to be too much, and she dropped the PhD program. She would have liked to get that degree, but it would not have advanced her career. The research she did for it ended up informing a training program she developed with the EPA for groups working on environmental contamination issues. She also applied her knowledge to coauthor and edit a book, *Community-Based Collaboration: Bridging Socio-Ecological Research and Practice* (University of Virginia Press, 2011).

As director of the GIC, Karen is at the forefront of mapping natural assets and crafting plans for their protection. GIC demonstrates how to map, evaluate, and restore green infrastructure through building models, research, teaching, and design. Karen works with local, regional, and state governments on dozens of projects. In Virginia, the GIC team adapted the state's model to create maps and plans for counties and regions. Then they built state models of the highest quality landscapes for New York, Arkansas, and South Carolina. Karen, who originally envisioned making a model and planning guide for every state, has had to accept that she doesn't have time to do it all. Still, toward that end, Island Press (2015) has published her national guide for green infrastructure planning, *Strategic Green Infrastructure Planning: A Multi-Scale Approach.*

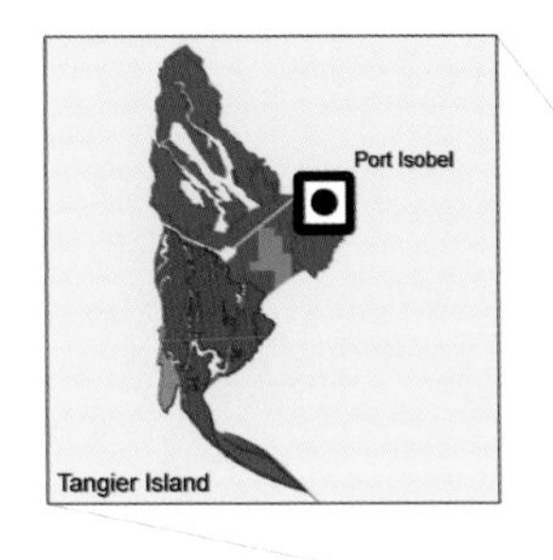

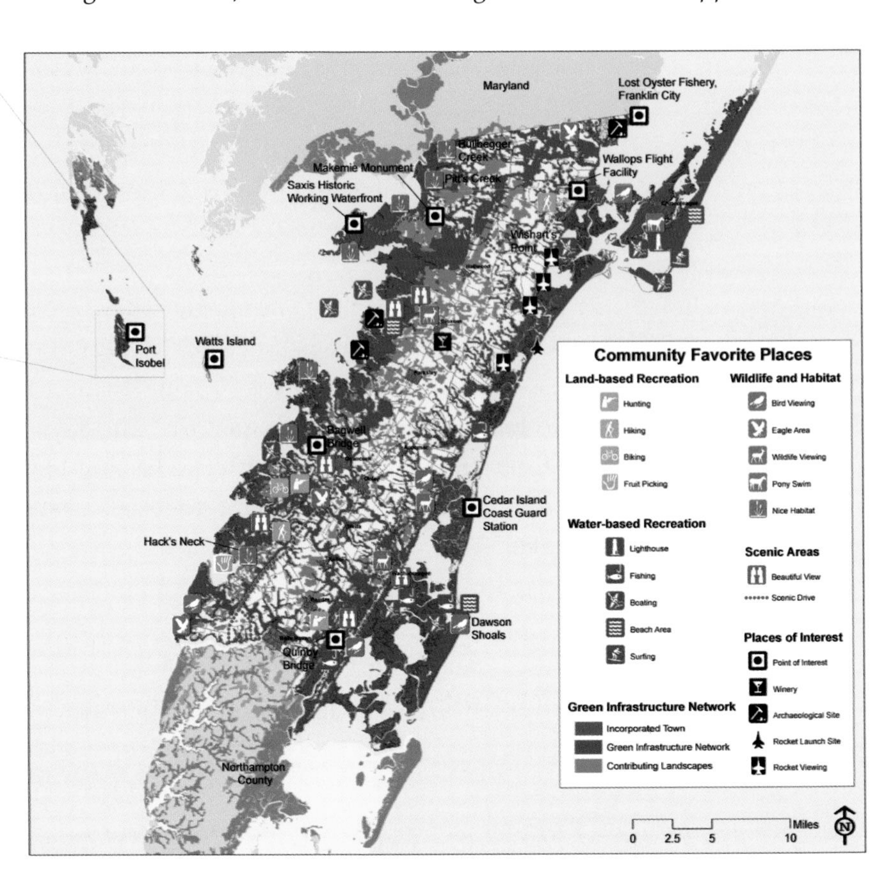

This map of Accomack County, Virginia, shows the green infrastructure network in green overlaid with favorite community resources. This map shows how the green infrastructure network supports many community uses such as walking, fishing, boating, and providing viewsheds.

When Esri wanted to build a national model to fulfill President Jack Dangermond's vision of a coast-to-coast green infrastructure network, Karen and her team provided their GIS script and tools that can run at any scale to help create a national model. She and her team worked with Esri's Applications Prototype Lab to work through the challenges of making a fast and reliable model for the

United States. "Esri has the reach and prominence to really put green infrastructure planning on the map (pun intended)," she says.

Throughout her career, Karen has often been either the youngest person or the only woman in the room, including serving on three federal advisory boards where she was the only woman. Growing up with three brothers prepared her to be comfortable in a male-oriented setting. She does not let herself be daunted; she just works harder. For instance, as part of her university studies, she took a groundwater hydrology course from the man who designed the cleanup plan for Three Mile Island, America's largest nuclear accident. When she arrived at the first class, she saw that she was the only undergrad among graduate students and practicing geologists. She studied and worked hard to understand the material and was able to match their performance in class.

Karen sees herself as a risk taker. Because she needs creative freedom to be happy, starting her own company is a perfect fit for her. Although supervising people is part of her job description, it is not her favorite task. "I just like the teamwork, not the boss part," she says. Her experiences and accomplishments have left her with the feeling that she can do anything she ultimately sets her mind to. Her goals for changing the world are as strong as ever.

I would like to normalize green infrastructure planning. Who would make a water supply plan without knowing their water recharge zones and aquifer capacities? Yet many communities plan with little data. I want everyone to see the value of this work and stop planning farm zones without asking where the good soils are, as one example. We have access to more data and processing speed than ever before, but we don't ask the right questions.

Karen wants every local government official around the world who is now sitting in a room making land-use and management decisions—many of whom are not looking at the right data to make those decisions—to begin their process with a map of their green infrastructure. And she wants them to use green infrastructure models to plan the future not only for their community, but for the planet's species. ✸

Kass Green

Providing the big picture on natural resources

THE VILLAGE ELDER BLEW THE SHORT, TRUMPET BEATS OF HIS HORN to call the village together. Kass Green had come to Ethiopia to analyze what kind of imagery could help secure property rights for farmers and their families working the land. In Ethiopia, all land is owned by the government, which grants individuals use rights only to farm the land—a system that provides the farmers and their families little security for inheriting or investing. Britain's Department for International Development (DFID), working to improve the incomes of the rural poor and enhance economic growth through rural land administration and second-level land certification, contracted with Kass Green & Associates to provide an independent analysis of the feasibility and appropriateness of using high-resolution satellite imagery for land registration activities. The goal is to provide policy makers and farmers with better documentation leading to better management of the land and land rights.

On this project Kass did what she loves and does best: working with resources and in the field to gather data that will inform decision-making so that policy makers can make better decisions.

Kass never expected that she would run a successful technology company, tour the world using GIS and imagery, and affect people's lives across the globe.

Even playing a small role in this important project to help people's lives was a thrill. Imagery and GIS give us the ability to definitively inventory and monitor resources worldwide. It is up to policy makers to decide how those resources should be allocated. I want to make sure that their decisions are based on accurate and complete information.

First, she had to get through first grade for the second time. "When you repeat first grade, you don't ever assume you're smart, you just keep working," she says. Kass learned to keep applying herself. She found out she was dyslexic, but that didn't discourage her; she kept studying. Her mother died when she was 14, forcing an abrupt move from the Midwest to California where she was raised by her aunt and uncle, who became like parents to her. Despite a chaotic home life, she was motivated to continue studying and working so that she could create her own life. She studied forestry in college and then natural resource policy in graduate school.

Kass's career has spanned a profusion of changes in technology as well as changes in how people approach certain issues. "When I started in this field, all the questions were about how much could be produced—how many cattle, how many board feet of lumber, how many bushels of corn, et cetera, which required a knowledge of economics. So I studied resource economics. But in the 1970s, people started to question if, how, and where those commodities should be produced—very different questions, which involved understanding the geographic location of resources, not just the amount of commodity produced. At the same time, GIS software was born, and Landsat imagery became available. So not only did we have the need to understand the *where* of resources, we also had the technology to do it. This is what sparked my passion for imagery and GIS," she says.

I morphed into a GIS/remote-sensing analyst because I discovered GIS and imagery to be much more explanatory and compelling tools for natural resource policy analysis than economics. However, I am very grateful for my economics training because economics is a powerful, although flawed, model of human behavior. Drawing on my economics background has always helped me in business decisions and policy analysis.

Kass studied forestry first, and then applied GIS and remote sensing to forestry issues before broadening to other applications such as agriculture and wildland fuels.

Kass says a strong grounding in a field such as meteorology, geology, ecology, urban planning, forestry, wildlife management, agronomy, and so on is critical to being able to employ GIS and remote-sensing technologies effectively.

Kass began her career first as a Washington, DC, lobbyist for the environmental movement and then at a forestry consulting and mapping company, Hammond, Jensen, Wallen & Associates (now part of Quantum Spatial). When one of the company's owners, who was also one of Kass's first mentors, Warren Halsey, asked her to use her economics background to evaluate the purchase of GIS software for the firm, she jumped at the opportunity. During this process she got to know another mentor, Russell Congalton, with whom she has collaborated for over 30 years on books, publications, and research. Russ was always patient, answering questions, and introduced Kass to the luminaries in the GIS and remote-sensing fields. "Warren taught me how to manage projects and negotiate. Russ taught me everything I know about GIS and remote sensing," Kass says. Kass explains that she has sought out mentors throughout her career, who always encouraged her questions, no matter how basic, so she could learn more.

In 1988, she and her husband, Gene Forsburg (also a forester), cofounded Pacific Meridian Resources, a GIS and remote-sensing firm that they grew to 75 employees in seven offices nationwide. "We used to joke that Gene probably would not have started a business on his own, and I would have started one, but it would have gone broke. It was the combination of our two personalities that made Pacific Meridian successful," Kass says.

Pacific Meridian worked with the US Forest Service, US Geological Survey (USGS), NASA, the National Park Service, multiple states, private timber companies, open space groups, and many more private and governmental clients on projects that combined geospatial analysis with imagery. When Kass began her career, most maps were derived from film aerial photography and manually etched on Mylar®, a clear, strong polyester film; nothing was digital. The first civilian digital satellite imagery, Landsat, was launched in 1972, and Pacific Meridian thrived using Landsat to map forests, endangered-species habitat, water use, crops, wildland fuels, wetlands, and land use. By the 2000s, aerial photography went digital, the first civilian high-resolution satellites were launched, and Landsat imagery became more accessible because of US policy changes. All the while, Kass incorporated these technological advances and appreciates what she has learned from all the employees, clients, and colleagues she has worked with over the years, all of whom shared their passion for their work and a desire to solve resource problems.

In GIS and remote sensing, there is always something new to learn—either a new application or a new technology.

Kass's focus has primarily been on land-use and land-cover mapping with an emphasis on wildlands. Her skill is to use the combined power of imagery and GIS to tease out and discover the relationships that determine the distribution of vegetation across the landscape. She has also been deeply involved in policy, especially Landsat policy. She is an enthusiastic proponent of the continuity of Landsat observations.

You can't manage what you don't measure. Landsat allows us to continually map and monitor the earth's resources.

In 2000, Kass and Gene sold Pacific Meridian to Space Imaging, which is now part of DigitalGlobe. Five different firms made offers to buy Pacific Meridian, which Kass found exciting and gratifying. She believes that the success of her business was because she and Gene followed their passions. After running half of Space Imaging for three years, Kass decided to focus her career on challenging and interesting remote-sensing mapping and policy projects and doing international consulting for organizations such as the Bill & Melinda Gates Foundation, the Omidyar Network, and DFID. Her mapping projects now tend to use high-resolution imagery but always in combination with Landsat and now also Sentinel imagery. Recently, she employed machine learning and object-oriented techniques to create detailed vegetation maps of Grand Canyon National Park, the national parks of Hawaii and American Samoa, and Sonoma County, California, using high-resolution optical imagery, lidar data, Landsat imagery, and multiple other datasets.

Probably some of the true highlights of my career have occurred when I am working in the field–whether it be agricultural fields in Africa, the forests of the Samoa, agency buildings in DC, vineyards and rangelands of Sonoma County, old-growth stands of the Pacific Northwest, coastal wetlands of Florida, or deep in the Grand Canyon. It is always an amazing experience to use imagery and GIS to study the dynamics of an ecosystem and to engage with the people whose culture and livelihoods depend upon it.

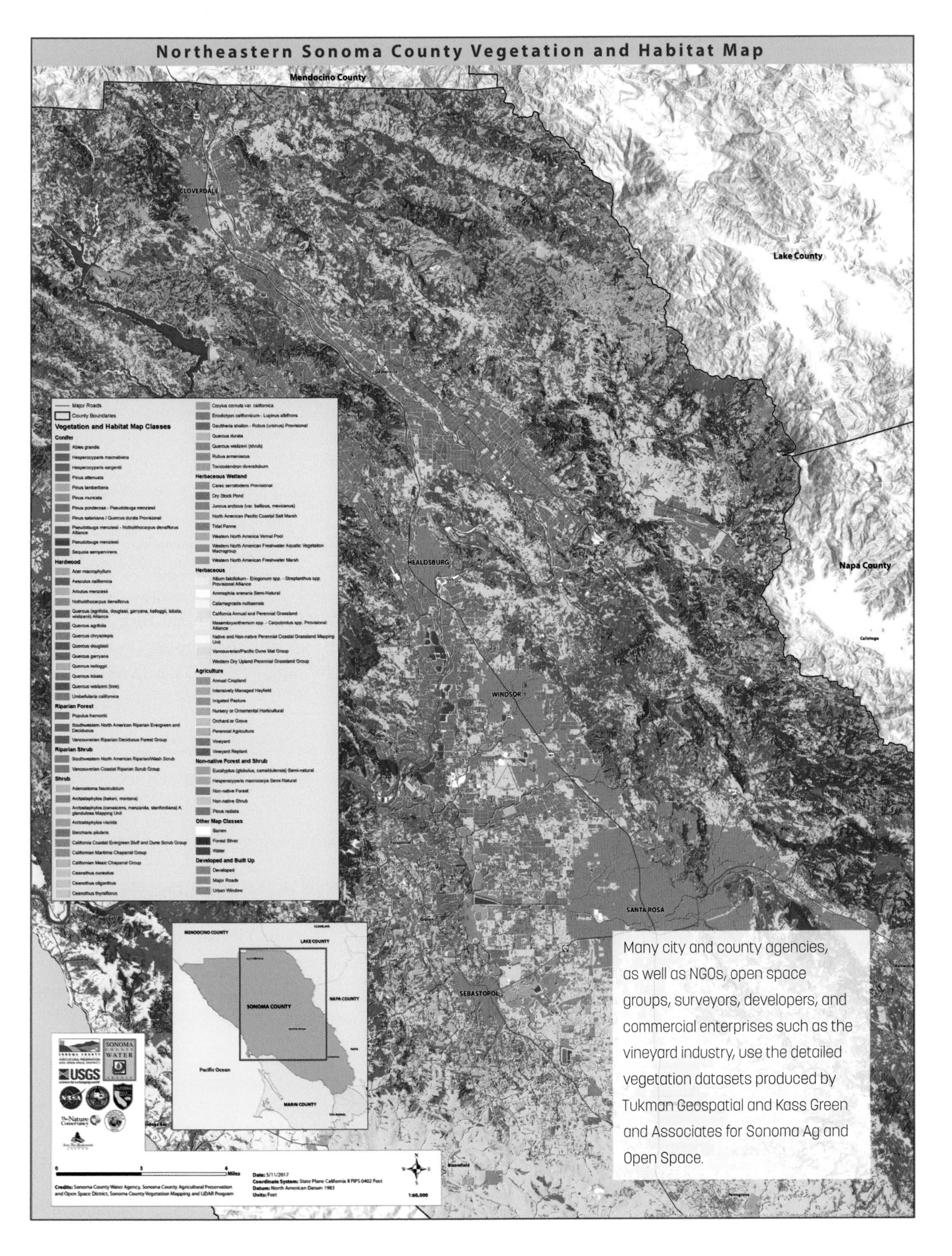

Many city and county agencies, as well as NGOs, open space groups, surveyors, developers, and commercial enterprises such as the vineyard industry, use the detailed vegetation datasets produced by Tukman Geospatial and Kass Green and Associates for Sonoma Ag and Open Space.

Kass calls *Imagery and GIS: Best Practices for Extracting Information from Imagery*, a book she wrote with coauthors Russell Congalton and Mark Tukman (Esri Press, 2017), the capstone of her career. "It has been a joy to capture all of our lessons learned from many years of GIS and remote-sensing projects and share them with others." She and Russ just completed the third edition of *Assessing the Accuracy of Remotely Sensed Data: Principles and Practices* (CRC Press, 2019). She and Mark continue to work with Sonoma County and will be monitoring the impact of the devastating wildfires that occurred in the autumn of 2017 with funding from NASA. They also have recently begun a project with the Golden Gate National Parks Conservancy to provide fine-scale vegetation data to other Bay Area counties.

Consistently committed to public service throughout her career, Kass was the first woman president of the Management Association of Private Photogrammetric Surveyors (MAPPS). She also served as president of the American Society for Photogrammetry and Remote Sensing (ASPRS) where she is a fellow and was honored as the first woman to receive its Honorary Lifetime Achievement Award, the highest tribute ASPRS bestows on any individual. Over the past 20-plus years, she has served on the University of California, Berkeley, College of Natural Resources Advisory Board, and on several federal advisory committees, for NOAA, NASA, and the Department of the Interior (DOI). From 2014 to 2018, Kass served as the chair of NASA's Applied Sciences Advisory Committee. As an ardent supporter of Landsat continuity, she served nine years on the National Land Satellite Archive Committee, testified before Congress, and cofounded the Landsat Advisory Group (a subcommittee of DOI's National Geospatial Advisory Committee), where she continues to serve today, providing advice on Landsat policy.

Kass offers advice to other women in pursuit of their careers:

I would tell this to all young women: Be bold. Be daring. Be honest. Be gracious. Be generous. Be passionate. Do not let anyone silence you or impede your progress, whether by bullying, body language, comments, or idle gossip. Find solid team members to work with in your career, and continually encourage their and your intellectual growth. Then listen, listen, listen.

Out of chaos comes opportunity. Do not be afraid to fail. Create a safety net of colleagues and systems so that any failure will not be catastrophic. You will learn from your failures. You are resilient.

She also says, "In choosing a mentor, colleague, or employee, choose those who are bright, passionate, and hardworking over those who are flashy and boastful. I would easily trade 10 self-proclaimed 'brilliant' colleagues for one who is industrious and enthusiastic.

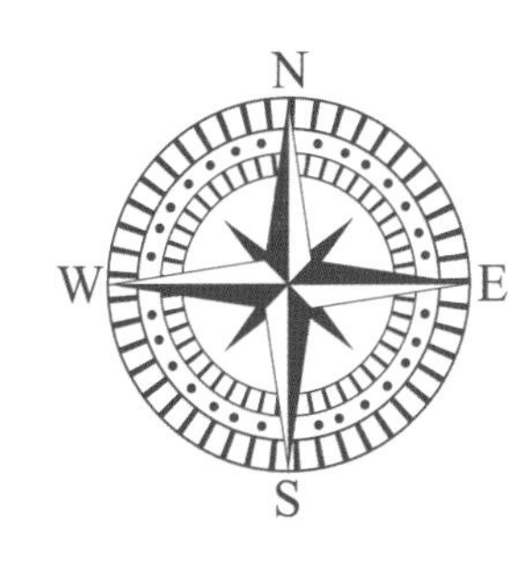

"Learn how to write well. If you can't write well, you can't communicate. And if you can't communicate, you can't succeed.

"In negotiating, don't poison the well. You might have to re-tread back through and need to drink that water someday.

"For anyone going into GIS and remote sensing, I would encourage them to first understand an application thoroughly, and then bring in the technologies to better understand that application." ✸

Kristen Kurland

The heart of a giving teacher

ANDREW CARNEGIE, FOUNDER OF CARNEGIE MELLON UNIVERSITY (CMU), famously said, "My heart is in the work." At the heart of Carnegie Professor Kristen Kurland's career and mission is teaching.

My students, who remain my major focus as well as my inspiration, benefit from my enthusiasm and willingness to serve as teacher, mentor, and role model.

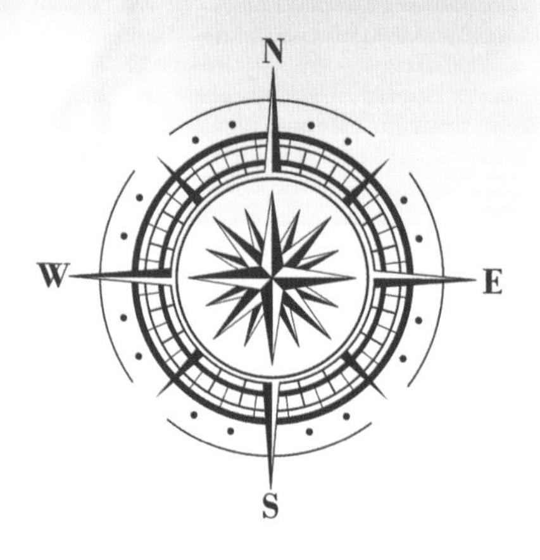

Even though Kristen wasn't especially fond of school as a student, she found her passion when she started to teach technology-based courses. At first, she taught other students and then professionals. She joined CMU's faculty in 1994 as part-time adjunct faculty and joined the full-time faculty in 1999. She estimates she has taught over 5,000 students over 25 years and mentored thousands of GIS projects in her classes. Kristen's teaching strategy and educational goals are to instill confidence, enthusiasm, and intellectual curiosity in her students. She promotes interdisciplinary collaborations in the classroom and the lab, continually develops new courses, innovates teaching tools and educational materials, integrates research topics in the classroom, and works to keep expanding her impact on teaching on a local, national, and even international scale.

Kristen's position at CMU has enabled numerous GIS projects through the years. One highlight was creating one of the first maps of childhood obesity. Working with endocrinologists and researchers at the University of Pittsburgh Medical Center (UPMC) Children's Hospital's Weight Management and Wellness Center in Pittsburgh, Pennsylvania, Kristen mapped young patients' changes in body mass index (BMI) over a five-year period and compared it with their proximity to green spaces and fast food. Their work was some of the first recognized by medical organizations such as the Pediatric Academic Society. Her research gained national attention, even receiving inquiries from former First Lady Michelle Obama's personal chef and director of her Let's Move initiative.

Growing up in western Pennsylvania as the youngest of four children, Kristen played hard outside in the summer swimming and riding bikes and in the winter went sled riding down her neighbor's long front lawn. She and her best friend, the boy next door, used their imaginations, transforming her front porch into a ship, putting on plays for friends and family, and making things from items they found.

Kristen jokingly says:

I'm convinced I created the prototype for the iPhone when I made an "anything thing" from a Tic Tac® box that I covered with paper and drew a radio, phone, and TV that I pretended were an all-in-one device.

Kristen's passion was drawing and art. Her mom signed her up for a summer art class with mostly adult students. "They humored me, telling me my drawings were as good as anyone's. One of my proudest moments was winning $75 in sixth grade for a contest for the best drawing that advertised upcoming events at a local mall. With my gift certificate, I bought a small gold watch because my mom told me it would be something I could own forever. To this day, I think of that prize every time I wear it."

Kristen's winning sixth-grade drawing, which is framed and hangs in her office.

Kristen was never encouraged to follow her passion for art in school. She was in a gifted academic program and was a straight-A student, but she wasn't really interested in school. "I mysteriously fell ill every Sunday in an attempt to stay home so I could play and create. Once in fourth grade, in a fashion rivaled only by Ferris Bueller, I was able to fake illness and stay home for almost two weeks by running the thermometer under hot water."

Kristen convinced the school board to allow her to shorten her high school experience by graduating after her junior year.

Intending to follow in her brother's footsteps, Kristen studied to become a doctor at the University of Pittsburgh. She majored in chemistry, worked at the medical library at Children's Hospital of Pittsburgh of UPMC five days a week, and volunteered on medical research projects, including working in the operating room during liver transplant surgeries performed by renowned surgeon Dr. Thomas E. Starzl.

Yet Kristen wasn't passionate about medicine and, in the second semester of her sophomore year, pleaded with her adviser to help her find another major. Her adviser realized that Kristen's courses, including math and physics, could be applied to a degree in architectural studies and that she could graduate as planned. She had only to take the core architecture classes, plus art, sculpture, and architectural history and theory.

Eureka! I could finally apply my love for art and not lose the science courses in which I'd worked so hard. And all's well that ends well. My job at the Children's Hospital of Pittsburgh medical library was not in vain, because that's where I met my husband whom I've been happily married to for nearly 27 years.

In college, she wasn't exposed to computers, except for editing her English papers on a Mac® computer in the basement of Pitt's iconic Cathedral of Learning. Kristen's college experience ended early too when she was offered a job working for a local engineering company that was partnered with a Dutch engineering firm. The

Though Kristen didn't realize her current passion or interests until after graduating from college, she always demonstrated optimism and good organizational skills. Here, she's getting ready to give a GIS talk for Grand Rounds at the University of Pittsburgh Medical Center headquarters.

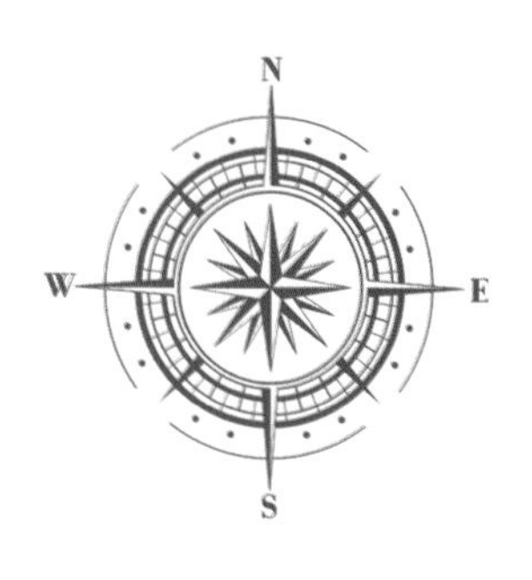

job required a brief move to the Netherlands to learn Autodesk® AutoCAD® software. Although she was considering graduate school in architecture and had filled out the forms for Carnegie Mellon, Kristen decided to put that on hold and take the challenge of learning a technology that was new to the field while getting a chance to see the world. In exchange for letting her leave school early, Pitt would give her course credit if she agreed to teach AutoCAD to fellow students after returning from the Netherlands. Even though her Dutch mentor didn't speak much English and the AutoCAD menus were partially in a foreign language, Kristen loved learning AutoCAD and found a real passion for working with computer-based technology.

In the early 1990s, Kristen worked for a Pittsburgh-based company, Computer Research Inc. (CRI) in technical support for AutoCAD, Computer Aided Facility Management (CAFM), and related software and as an instructor, teaching many of the architecture and engineering firms in Pittsburgh how to implement these applications. CRI closed its CAD division while she was on her honeymoon, and when she returned, she started her own consulting firm. She worked with architects, engineers, large organizations such as Westinghouse Electric Corporation, federal government agencies (she held a top-secret clearance for many years), hospitals, and local governments. Her first exposure to GIS was in the early 1990s when she helped Pittsburgh's City Planning Department and Allegheny County create Pittsburgh Allegheny GIS (PAGIS) by developing base layers for the city and county.

In 1994, Wil Gorr, a professor at CMU, recruited Kristen to teach AutoCAD to at-risk youths in Pittsburgh for a mentorship program he started called InfoLink. InfoLink was a summer-long program for teaching high school juniors and seniors technologies such as AutoCAD, GIS, and web design, and then placing them in paid internships with local architects, engineers, and other firms. Impressed by her teaching in InfoLink, Wil enlisted Kristen as an adjunct professor at CMU. Kristen continued to work with her consultancy firm until 1999 when she became full-time faculty at Carnegie Mellon. Today she is a professor in the School of Architecture and the H. John Heinz III College of Information Systems and Public Policy. She also holds a courtesy faculty appointment in Civil and Environmental Engineering. Much of her research and teaching at CMU is GIS-related, but she also teaches AutoCAD 2D/3D, CAFM, BIM, and other related spatial technologies. At Heinz College, she teaches a course to physician executives in the Master of Medical Management program called Enterprise Data Analytics, which shows how geospatial projects can relate to medicine.

Wil, pictured with Kristen at Carnegie Mellon's presidential inauguration in 2013, has been Kristen's main career mentor.

Kristen credits Wil with showing her how to successfully navigate teaching and research at CMU, how to think and write critically, but most importantly, how to give back to students and society. She says, "Wil's kindness and humor showed me that one can be successful and not lose sight of what is important in life. By example, Wil has helped me develop a great life/work balance."

While working in the InfoLink program, Kristen and Wil began to write GIS tutorials together. They have continued to be coauthors for Esri Press, starting the GIS Tutorial workbook series, coauthoring *GIS Tutorial 1 for ArcGIS Pro: A Platform Workbook* (2017) and editions of *GIS Tutorial for Health*, and just finishing a new edition of *GIS Tutorial for Crime Analysis* (2018). "Data and examples in our books draw upon [our] experience in both the classroom and with projects that Wil and I have [worked on] in our combined over 75 years [of] teaching and research. The roots of many of my health-related projects and collaborations began with our *GIS Tutorial for Health* workbook [2006]."

Other pioneering work included childhood lead studies (with the Allegheny County Health Department), pedestrian accidents (with trauma surgeons at Children's Hospital of Pittsburgh), asthma-related research (with physicians and fellows in the Pulmonary Division at Children's Hospital of Pittsburgh), and analysis of travel time to hospitals (with critical-care physicians and researchers at UPMC). Work with the Pittsburgh Police, Department of City Planning, and numerous nonprofit and other firms also contributes to examples in their core GIS Tutorial series and *GIS Tutorial for Crime Analysis* workbooks.

For another project with Heinz College Health Care Policy and Management students and Children's Hospital of Pittsburgh, on which Kristen served as faculty adviser, the hospital wanted to know more about patients who came to the emergency department yet had low-acuity (nonurgent) reasons for their visits. Those nonthreatening problems could be treated in the outside community in urgent-care centers. Avoidable emergency room visits are costly, and low-acuity visits take valuable time away from high-acuity patients. Kristen's students identified specific neighborhoods and public housing locations and worked with faculty with urban planning and design expertise to help identify why these neighborhoods were problematic (for example, relocation because of gentrification issues) and offered potential education and policy solutions. "The nature of interdisciplinary work at Carnegie Mellon, plus our relationship with local government and others, not only allow us to identify problems, but also provide mechanisms to change policy."

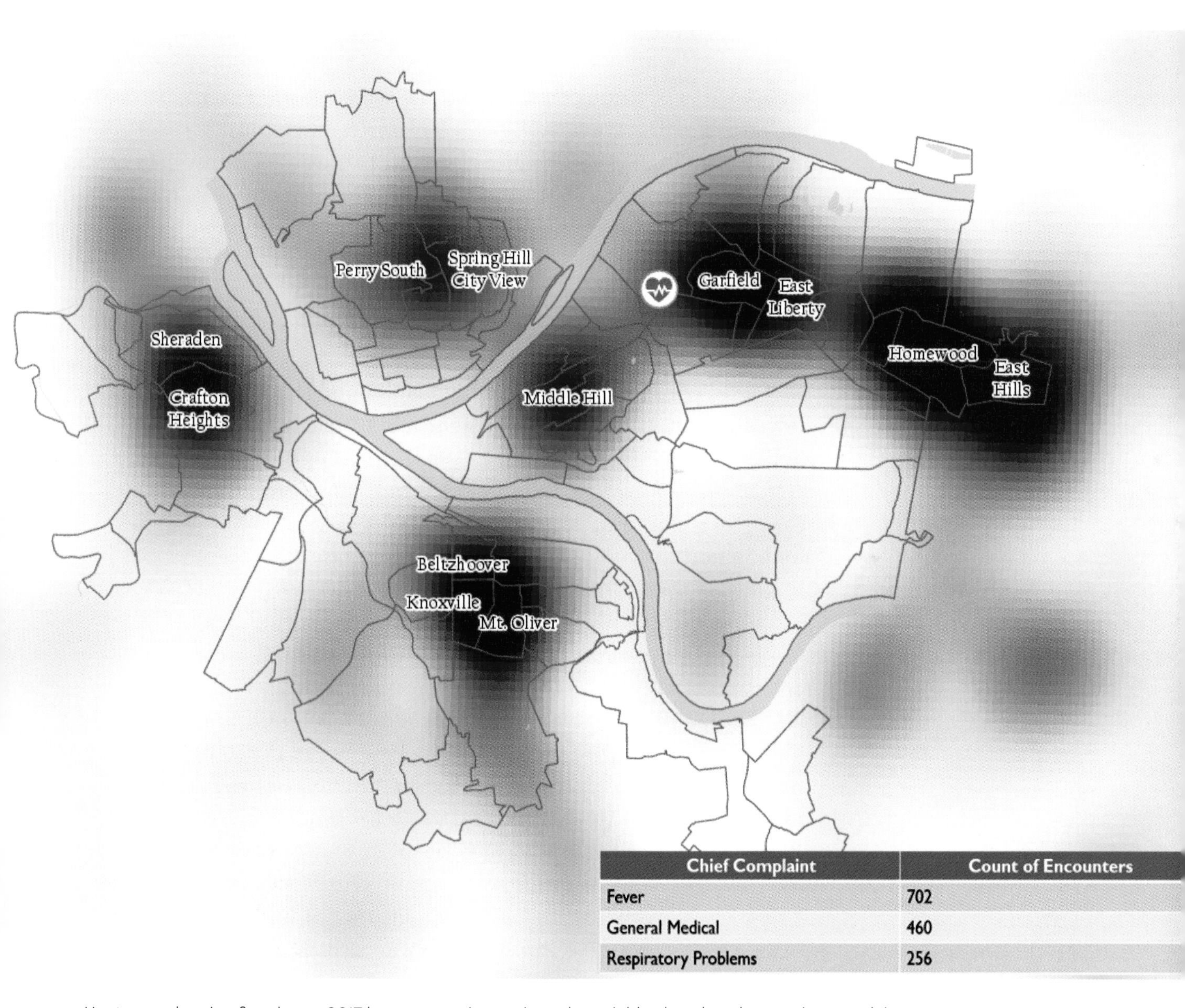

Chief Complaint	Count of Encounters
Fever	702
General Medical	460
Respiratory Problems	256

Heat map showing fiscal year 2017 hot spots using patients by neighborhood to show major complaints.

Kristen demonstrates her desire to give back and pay it forward in her commitment to supporting education at CMU. She makes financial donations with the goal to help provide funds to students who can't afford the university. For nine years, she served on the Executive Board of CMU's donor recognition society, Andrew Carnegie Society (ACS). She was president of ACS from 2014 to 2016 while serving as a university trustee.

Currently, Kristen is on the distribution committee of the McCune Foundation, one of only three community members sitting with the other four members from the McCune family. The McCune Foundation donates to organizations in southwestern Pennsylvania with the goal of making the region a better place to live. Kristen used GIS to help both the ACS and the McCune Foundation better understand donors, grant recipients, and communities affected by grants and donations.

As trustee at Carnegie Mellon, Kristen helped write the university's current strategic plan.

It's clear that the cost of higher education will soon exceed an amount that many can afford. Colleges and universities will need to reinvent how they educate students while still upholding academic rigor through research and scholarship.

Kristen works with Metro21, a CMU interdisciplinary institute that deploys research and development done at CMU to the Pittsburgh region. Her current project uses 3D visualization for urban planning and demonstrates how faculty and students can test and evaluate GIS and virtual reality tools for the city to consider adopting.

Kristen continues to innovate in the classroom, always learning and sharing what she learns.

Kristen teaching students in the School of Architecture the importance of considering health in urban design.

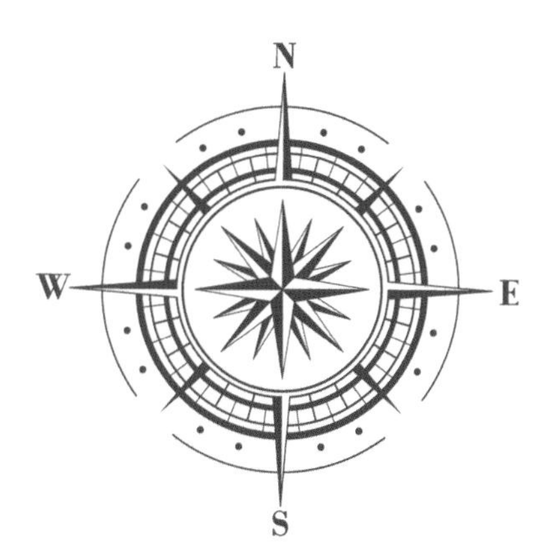

No matter what you study, your options remain open. STEM fields or, better, STEAM, adding ["A" for the] "Arts," are no longer silos. Engineers work with planners and citizens to develop smart technologies used in cities, architects work with designers and gaming experts to inform the public of changes to the landscape, [and] computer scientists in robotics work with mayors and policy makers in cities to deploy driverless vehicles. The list is endless.

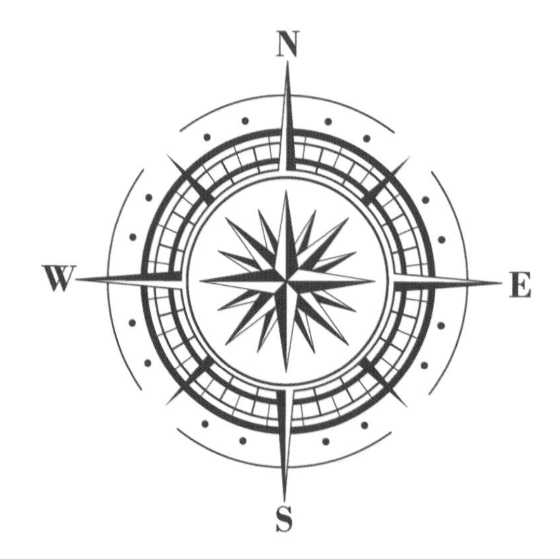

Kristen with David, *center*, his wife, Judi, *left*, and Kristen's husband, Geoff, getting ready for a Pittsburgh Steelers football game.

Kristen has also been influenced by David Lewis, whom she met at Carnegie Mellon after he was exiled from South Africa in the early 1940s for defending a nonwhite student who was attacked while jogging on campus at night. For many years, Kristen worked closely with David on urban design projects and learned from him the value of engaging citizens and their communities. In addition, David taught her that courage and following one's passions such as sports (David and Kristen are both huge hockey and football fans) and art can be integral parts of one's private and professional lives. Coincidentally, David's daughter, An Lewis, teaches GIS at Kristen's alma mater, the University of Pittsburgh, showing how life truly does come full circle. ✸

Nancy La Vigne

Using quantitative research to tip the scales of justice

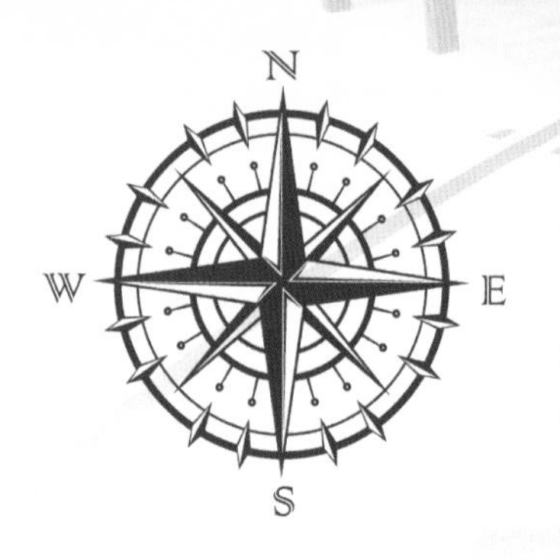

IT TOOK NANCY LA VIGNE A WHILE TO FIGURE OUT WHAT SHE WANTED TO do with her life. When she was in elementary school, she excelled in math, but lacking encouragement from teachers or peers, her interest waned. In the second and third grades, she loved poetry, and a caring teacher encouraged her to pursue creativity rather than math and science. Geography would end up being a big part of her career, but in middle school, she hated the subject because a teacher assumed she knew nothing about the topic. In high school, she was inspired by an advanced chemistry teacher and thought she might major in chemistry. Looking back, she says, "It's astounding when you consider how much teachers can influence your love—or disdain—for a given subject!"

One thing she knew for sure: she would be pursuing her education. Her mother, a first-generation Italian American, was a strong believer in the value of education and was the first in her family to earn a college degree; she ultimately acquired a master's degree in education. Nancy's parents were both educators, having met overseas while teaching children of American servicemen stationed there. When they returned stateside, they began a family in a New Jersey suburb of New York City, with Nancy the youngest of three children. Both of Nancy's parents worked full time, and her mother inspired in her a strong work ethic and a belief that she could accomplish any goal she put her mind to if she gave it her best effort.

Left, Nancy (third from left) with fellow high school seniors in a chemistry class. *Right*, at a recent class reunion, Nancy met up with her high school teacher Lorena Tyson.

Nancy veered toward the social sciences at Smith College in Northampton, Massachusetts, a small, all-female institution in the classic liberal arts tradition.

Smith was a great place to explore interests, and because it was all women, there was a freedom to inquire and express ourselves in the classroom without being dominated by, or drowned out by, men. That said, back then Smith had no premajor requirements; while we were encouraged to explore new subjects and disciplines, someone could easily graduate Smith having never taken a single math course. I was one of those students, and looking back, I regret the missed opportunity to explore and advance my quantitative aptitude while in undergrad.

Nancy graduated from Smith with a degree in political science and economics, and then worked as an intern in a congressional office on Capitol Hill. She saw a job opening as a legislative assistant at a nonprofit gun control advocacy organization, and after she took that job, she met victims, law enforcement officers, and members of the public, all of whom cared deeply about public safety and had a strong case to make for policy change. This experience was Nancy's aha moment: she realized that criminal justice policy in the late 1980s was driven by emotions and politics. She understood that this was a field where she could make a difference by generating empirical research to elevate the policy conversation to a level of discourse guided by facts, rather than beliefs and biases. Now she had found what she wanted to do.

In graduate school at the LBJ School of Public Affairs at the University of Texas at Austin, Nancy began to develop an interest in statistics, analytic decision-making, and multivariate analysis. After earning her master's degree in public affairs, she worked in Texas as director of research for a sentencing commission in the early 1990s. She wanted to examine the degree to which judges' sentences were racially biased. She was told very publicly by a man who had a PhD that she didn't know how to do that analysis and that, in fact, it wasn't even possible. Nancy knew it was possible and that this person did not want her to pursue the research for political reasons. Still, the damage was done: he had undercut Nancy's credibility. It was then that she decided she must pursue a PhD if she was going to be taken seriously.

Continuing to advance her quantitative skills, Nancy worked toward her PhD in criminal justice at Rutgers, the State University of New Jersey. She became interested in the role of place in crime and criminal behavior and decided to focus on the spatial analysis of crime. She used GIS to organize and analyze data such as the volume and types of crimes occurring in the aboveground areas within a one-mile radius of subway stations. By the time she completed her doctorate degree, she was well versed in both quantitative and qualitative methods and proudly describes herself as a "mixed methods" researcher. She sees the value of both forms of information and evidence.

Nancy, *at table center, lower right,* leading a group of community members and law enforcement officers in a discussion of community perceptions of police bias.

Nancy became the founding director of the Crime Mapping Research Center with the National Institute of Justice, US Department of Justice (DOJ). The center was dedicated to helping practitioners and researchers employ and gain value from GIS as a tool for understanding and addressing patterns of crime and criminal behavior. At the DOJ, Nancy found an important mentor in her professional life, Dr. Pamela Lattimore. Pamela, who gave Nancy the opportunity to establish and lead the Crime Mapping Research Center, "is a great example of a mentor who was not concerned with taking credit but rather supporting the professional advancement and accomplishments of others," Nancy says.

I sometimes wonder what my career would have looked like without female role models like Dr. Lattimore.

In 2001, Nancy then began working for the Urban Institute, a nonprofit research institute in Washington, DC, that, like Nancy, believes that decisions shaped by facts, rather than ideology, have the power to improve public policy and practice. As a senior research associate, she developed and led the Reentry Mapping Network, a collection of a dozen jurisdictions engaged in mapping the locations of people exiting prison to understand their needs for services and treatment in support of the reintegration process. Nancy is now vice president of justice policy at the Urban Institute, where she leads a staff of over 50 engaged in research and evaluation across a wide array of crime and justice topics. Her work consists of a challenging balance of setting strategic direction, managing people, leading her own research, and being the public face of the institute on topics of criminal justice.

It's not always easy, and I am often asked why on earth I want this job! But I find that having different roles within the same position is energizing; I am constantly encountering new opportunities to learn and grow.

Her own research has spanned many topics, including understanding the challenges of people exiting prison and returning to their communities, the impact of public surveillance cameras on crime, and the perceptions and experiences of people residing in the highest crime and most heavily policed communities regarding policing and public safety. A common thread among these and most of her other research projects is the importance of place and spatial context—GIS has played a significant role in her research.

Working in a research institute rather than an academic setting suits Nancy because she wants to influence people who are positioned to change criminal justice policy. She wants to conduct research on the patterns and impacts of crime and policing that can promote reforms that enhance public safety while reducing racially biased outcomes.

Nancy, at an Urban Institute event, interviews Ronald R. Davis, former director of the Office of Community Oriented Policing Services, on how research can inform efforts to improve police-community relations.

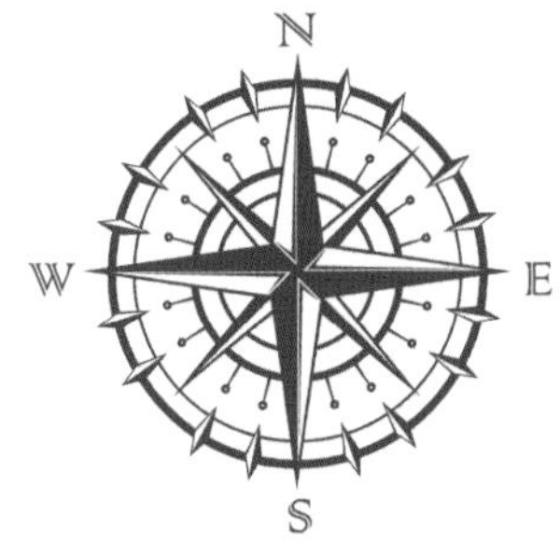

Although it is important to share this research with policy makers, Nancy and her colleagues are also striving to bring their findings to the people and communities who are most affected by it. A recent example is a project Nancy led that interviewed community members in the highest crime, most heavily policed communities about their views of public safety and police bias. That project engaged community members at every step of the process, from designing the survey to interpreting the findings to developing policy recommendations to share with local law enforcement. The project was rewarding but resource intensive, involving door-to-door interviews with community members. Nancy sees advances in information technology and the ubiquity of smartphones as mechanisms that will facilitate the collection of spatially tagged data in the future and democratize and empower communities to promote change from the bottom up.

Rising to become a leader in her field, Nancy recognizes that her career success has benefited from both male and female teachers and mentors. However, women have played a markedly strong role.

Without female role models, I doubt I would have accomplished as much as I have. I suppose that's why the gap continues to exist today; it takes time for diversity across all dimensions to reach all levels. Until then, we need to be more intentional in providing support to underrepresented populations of all kinds.

Nancy encourages young women to seek out mentors who will help them pursue their passions and promote their professional advancement.

A lot of younger professionals have the notion that mentors will find them, or that their direct supervisors are supposed to serve in that role. That hasn't been my experience. I've found that the best mentors are ones I have proactively sought out, based on people I admire and aspire to emulate.

She also recognizes that she is in a unique position to pay it forward. "I welcome opportunities to share my experiences and advice with those who are starting out in the field. Despite a busy schedule and competing demands, it's important to me to make time for those relationships."

However, this leadership role is also one of her biggest challenges.

The prominence Nancy has achieved in her field and the ease with which she can be found online has meant that managing her inbox drives her crazy. "My biggest burden is the never-ending requests and demands that mostly appear via email. I simply cannot keep up, and then I am plagued with guilt when I discover that I have left a request unanswered." Most of these emails come from potential partners in the criminal justice field and, even more importantly, from students and young professionals.

I aspire to make time for all of them, but it's not always possible. Yet I recognize that it's an important responsibility to attend to these requests and nurture future leaders.

In thanks to all the mentors who have made time for her, Nancy now strives to reciprocate by passing on a legacy to the next generation. ✸

Wangari Maathai

The power of one little hummingbird

LAND THAT USED TO BE DRY AND BARREN NOW STRETCHES LUSH AND verdant across thousands of miles of Africa, expanding like a long green belt. Before Wangari Maathai planted her first tree in Kenya, nothing much was growing there anymore. Decades later, 50 million trees cross the land and reach for the sky, and all because of a hummingbird.

A popular African children's story now doubles as a parable of one heroic life: the tale tells of a huge fire burning in a massive forest, while the forest animals stand frozen in a clearing watching it. They feel overwhelmed by the situation, hopeless and powerless. Only one creature decides to do something about it. A little hummingbird flies to the stream, picks up a drop of water in its tiny beak, flies back to deposit it over the blazing flames, and then does the drop all over again. Buzzing back and forth, over and over. Shaking their heads, the animals—including the elephant whose trunk could carry much more water—say to the hummingbird, "What can you do? You're too small to put out a fire that big. Your beak can only bring one tiny drop of water at a time." Undeterred, and undiscouraged, the hummingbird continues to douse the flames, drop by drop, saying, "I am doing the best I can."

That hummingbird's stance of doing all it could in the face of a greater need was the mantra of a strong-minded woman who motivated millions of others to take their first steps out of poverty, brought about by deforestation and environmental degradation,

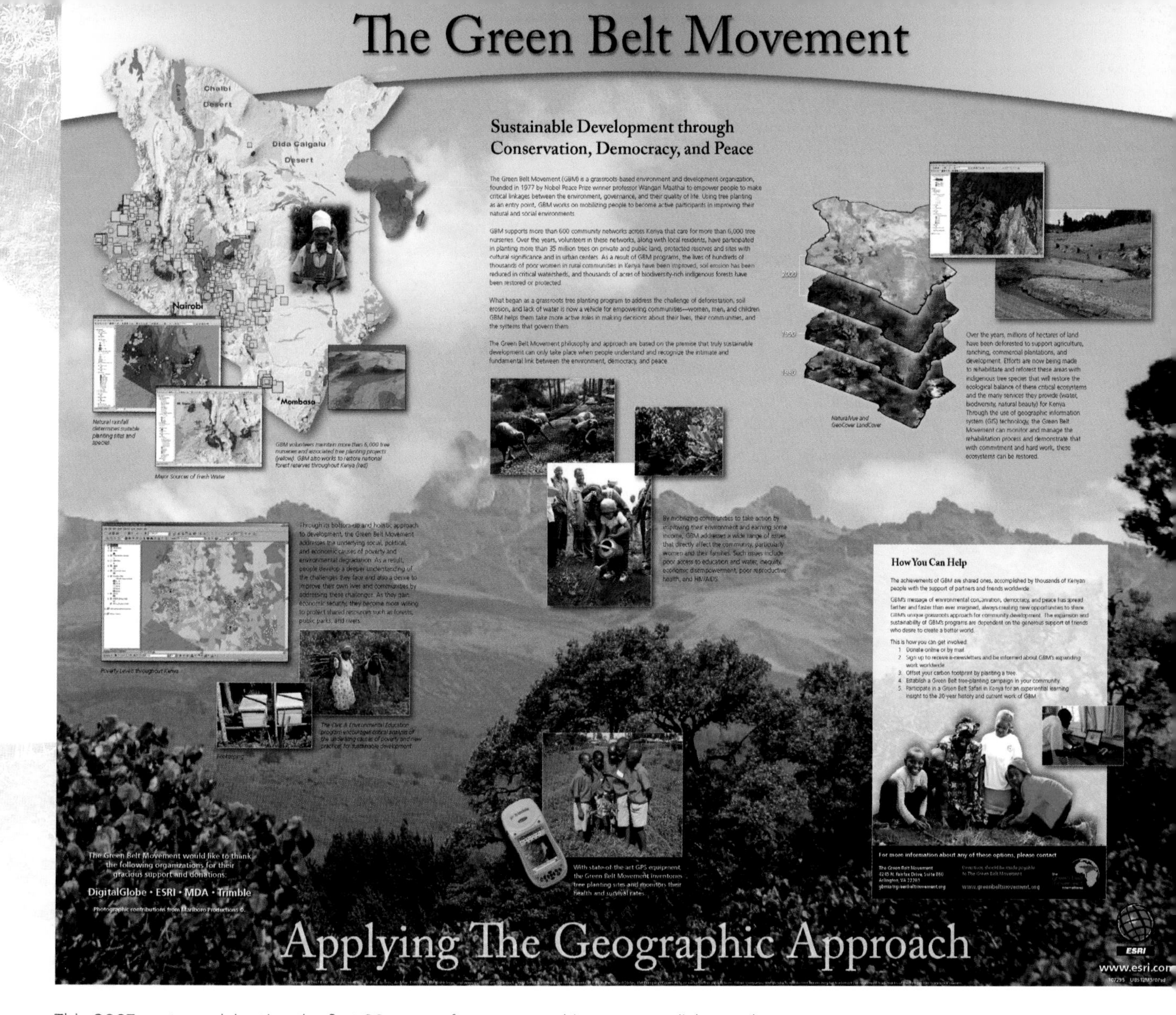

This 2007 poster, celebrating the first 30 years of grassroots-driven accomplishment by the Green Belt Movement, shows that many lives have been improved and many acres of indigenous forest have been restored or protected.

by planting trees and seeding a brighter future. The first African woman awarded the Nobel Peace Prize, in 2004, and also the first environmentalist to be so recognized, Wangari Muta Maathai saw the connections—between planting trees and planting the seeds of resilience and democracy. Trees restore water to the land and provide shelter, fuel, and a livelihood. Women planting trees, and so earning a living by their work, helps promote women's rights and, in turn, human rights.

Courtesy of GBM.

You cannot enslave a mind that knows itself. That values itself. That understands itself. When you know who you are, you are free.

Organizing Kenyan women in local communities to plant indigenous trees grew into a grassroots movement with a global effect. If they could plant one tree, they could plant many more trees, even a million more. Thus, planting nine seedlings in her backyard on World Environment Day in 1977, Wangari founded the Green Belt Movement (GBM), which opened the eyes of the world to the relationship between sustainable use of natural resources and emancipation from an environmentally disruptive government. And it helped lay the groundwork for GIS to take hold decades later with a lab to chart the precise placement of trees across Africa in spots where they are needed most.

I started planting trees and found myself in the forefront of fighting for the restoration of democracy in my country. The environment is not an issue for tomorrow, it is [an] everyday issue. It is the air we breathe, the water we drink, and the food we eat, and we can't live without these things. All over the world people are fighting over water, over food. ... We plant the seeds of peace.

Wangari upheld the spirit of her people and of the land. Courtesy of GBM.

Every person who has ever achieved anything has been knocked down many times, but all of them picked themselves up and kept going, and that is what I have always tried to do.

Wangari was born in a poor, rural area of Kenya called Nyeri in 1940, when the women water gatherers could still find clean water close by and when each tree had a spirit, and even a personality. Wangari has described her childhood this way: "I would visit the stream near our home to fetch water for my mother. I would drink water straight from the stream. Playing among the arrowroots, I tried in vain to pick up the strands of frog eggs, believing they were beads with which I could adorn myself. But every time I'd put my little finger under these beads, they would break. Later I saw thousands of tadpoles, black, energetic, wiggling through the clear water."

This memory had a special meaning for Wangari.

This is the water I inherited from my mother. Today, 50 years later, my stream has dried up, women walk longer distances to fetch water that is not always clean, and children may never play with the tadpoles and the frog eggs, and they may never know what they lost. Our challenge is to restore this home for the tadpoles and to give back to the children.

Winning a Kennedy Scholarship to study in the United States, Wangari earned a degree in biological sciences from Mount St. Scholastica College in Kansas in 1964 and a master of science degree from the University of Pittsburgh in 1966. Upon returning home from college, Wangari was shocked by what she saw when she looked down from the plane—the Africa that deforestation had rendered. The second-largest continent in the world lay cut and desiccated by exploitation of its natural resources. Back in her village, she heard the situation on the ground from women she knew in the community; they complained about the poverty, the struggle, and

The 1960 Kennedy Scholarships program sent 300 Kenyans to study at US colleges and universities, including Wangari at 20. Courtesy of GBM.

the strife. Wangari says "Why not plant a tree?" was the first thought that came to her.

In her Nobel lecture decades later, Wangari explained why this simple thought was such a good idea.

> **Tree planting became a natural choice to address some of the initial basic needs: it's simple, attainable, and guarantees quick, successful results within a reasonable amount of time. Those factors are important to sustain interest and commitment.**

Wangari helped nurture women even as she helped them learn to nurture trees. Courtesy of GBM.

When we plant trees, we plant the seeds of peace and the seeds of hope.

The GBM trained the women how to grow trees, creating jobs for them even as they worked to improve the community's natural resources. Through this simple action, the women of Kenya grew to feel empowered.

Wangari's stance:

The state of the environment is a reflection of the governance in place. Without good governance, we cannot have peace.

Use of the tree as a symbol of peace was in keeping with the widespread African tradition of using a staff (made from a tree branch) to signal conflict resolution. During territorial disputes, for example, a staff was placed between the two parties, a signal to stop fighting and seek peace.

In 1992, in the face of governmental privatization of the country's resources, the GBM, originally mobilized to plant trees, became a people's movement mobilized to protect trees. Eventually, Kenyan public demand for democratic governance prevailed. Planting trees together as a people had helped cultivate peace in Kenya.

It is the people who must save the environment. It is the people who must make their leaders change. We must stand up for what we believe in. And we cannot be intimidated.

In 2004, Wangari was awarded the Nobel Peace Prize. In *Wangari's Trees of Peace: A True Story from Africa* (Harcourt, 2008), author Jeanette Winter tells children what Wangari had accomplished by then: "By 2004, 30 million trees had been planted, 6,000 nurseries existed in Kenya, the income of 80,000 people had been increased, and the movement had spread to 30 African countries—and beyond." On the day she was notified of winning the Nobel Prize, Wangari made her way to the base of Mount Kenya and planted a Nandi flame tree.

In her Nobel acceptance speech, Wangari, holding her Nobel medal, said, "We are called to assist the earth to heal her wounds and, in the process, heal our own–indeed, to embrace the whole creation in all its diversity, beauty, and wonder." Courtesy of GBM.

Wangari, who died in 2011 at the age of 71, has left a legacy of courage and perseverance. Today, one of Wangari's three children, Wanjira Mathai (yes, her name is spelled correctly), carries on her mother's work, fighting climate change. She chairs the GBM and is the cochair of the World Resources Institute's Global Restoration Council. Echoing her mother's sentiments, Wanjira points out, "She never did anything to please or because she wanted to be popular. She did everything she did because she always felt it was the right thing to do. That was a constant in her life."

Wangari and daughter Wanjira, *right*. Courtesy of GBM.

Young women who continue in Wangari's footsteps are aided by the scholarship fund in her name at the Wangari Maathai Foundation (WMF). In her role as chair of the WMF, Wanjira promotes a values-based education via its Hummingbird Program, to counteract what she says are the top environmental problems: human selfishness, apathy, and greed.

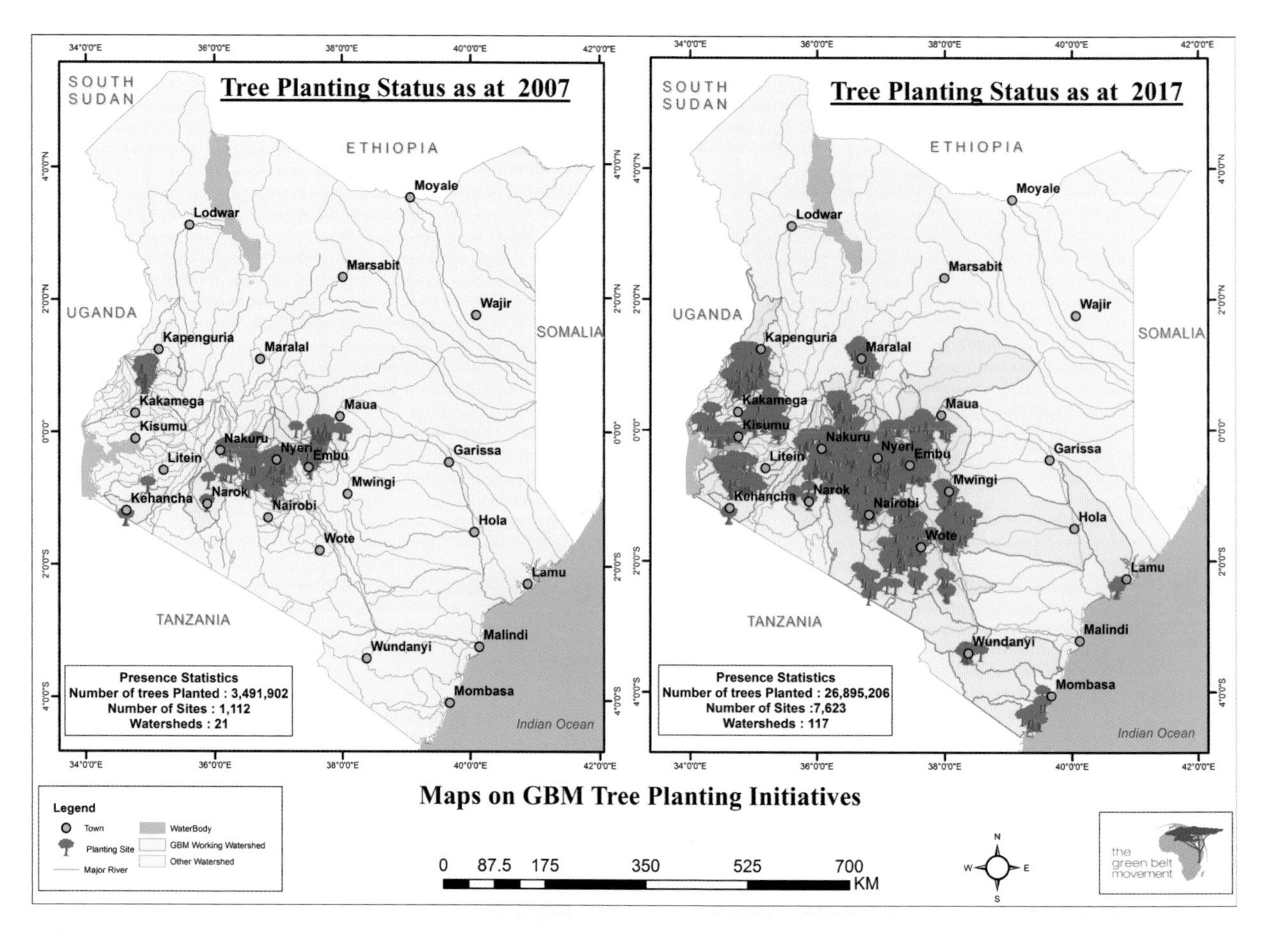

In her wisdom, Wangari created a movement that is still growing strong without her. In 2017, thanks to the GBM and those who carry on the work she started, 23 million more trees were planted than 10 years earlier. Maps by Nancy Neema, GIS officer, courtesy of GBM.

Through her life's work, Wangari showed people all over the world the value of an African parable—how even a little hummingbird can make a big difference. In her lifetime, Dr. Maathai wrote four books, served as a board member of 18 major organizations, and received 50 international awards along with 15 honorary doctorates. In 1971, the same year her daughter Wanjira was born, Professor Maathai became the first woman in East and Central Africa to earn a doctorate degree at the University of Nairobi where, five years later, she served as the first woman chair of its Department of Veterinary Anatomy and Physiology. In partnership with the university 34 years after that, she founded the Wangari Maathai Institute for Peace and Environmental Studies (WMI).

Trailblazer Wangari did all in her power to form an environmental greenbelt across nations and contribute to a greener, cleaner planet. And now, GIS software helps perpetuate the hummingbird effect through the scientific placement of trees.

Peter Ndunda, GIS Specialist and graduate of the University of Redlands who helped GBM establish its state-of-the-art lab nearly a dozen years ago to support 600-plus community networks across Kenya and the care of more than 4,000 tree nurseries, says GIS brought "a new way of thinking, planning, project monitoring, and creating solutions for sustainable development."

As Peter puts it, "GIS has brought the ability to answer the 'where, how, what, and why' in GBM—what kind of tree species to plant and where to plant them, what is the extent of deforestation in the five major mountainous forests in the country, and why some areas are more critical than others."

Originally, about 12 percent of the surface of Kenya was covered by closed-canopy forest. Today, it is down to only 2 percent.

"The United Nations recommends that a nation have a minimum of 10 percent forest cover to deliver the vital ecosystem services these forests supply to support sustainable development, such as fresh water for its people, agriculture, and wildlife," Peter says. "The reforesting needed to achieve 10 percent cover in Kenya is enormous."

And so was Wangari's commitment to help make it happen. ✸

Holley Moyes

Archaeologist explores Maya ritual caves

IN GUATEMALA ON THE SHORTEST NIGHT OF THE YEAR IN THE NORTHERN Hemisphere, a young dental hygienist in her mid-30s climbed up and scampered down the architectural remains of temple steps built close to 2,000 years ago. Because the guards at the site were on strike, she had this amazing setting almost all to herself. The moon rose full that night, and Holley Moyes sat atop Temple 4, the highest point at the site, used as the rebel base from which Luke Skywalker and Han Solo launched their attack in *The Empire Strikes Back*. In the ghostly moonlight, the thick tree canopy spread around her and howler monkeys chanting in the distance, Holley imagined herself a witness to the voices of the ancient Maya calling across time. In hindsight, Holley realized how much this experience affected her as she would eventually become a Mayanist, forever intrigued by this long-ago culture.

Holley Moyes is a cave archaeologist who primarily studies ancient Maya ritual cave sites. In Mesoamerica (and elsewhere), deep caves with dark zones were used almost exclusively as ritual spaces. Holley reconstructs ancient religious practices and correlates religious ritual with social and environmental events.

Holley's journey to becoming a cave archaeologist was circuitous, with elements of her history and education intertwined with opportunity and accident.

I have never been one to plan ahead too much and have tended to open the door when opportunity knocked. I know that living this way would drive a lot of people crazy, but I thrive on this somewhat chaotic existence and have trusted the universe to put me where I've needed to be when I needed to be there.

At the start of her career path, Holley followed her dream to be an actress. In high school, she spent as much time as she could in theaters and on stage. At Florida State University, she majored in theater, and then went to a private acting school in New York City. After finishing the acting program, she joined with theater friends to open a small acting company called the Neighborhood Group Theater. The members of the theater company were dedicated and hardworking and put on low-budget productions. New York was an exciting place to be, but the company had too few seats to sustain salaries, and Holley always had to find and maintain at least one outside job.

When Holley turned 30, she knew she had to do something else. She left the high rents of New York and found herself in a funk on her parents' couch in Florida, trying to figure out her next move. Her father, a dentist, convinced Holley to work in his dental office as an office manager. Her father's dream was for Holley to become a dental hygienist, and Holley didn't have a better idea. So she went to dental hygiene school but knew as soon as she started that it

Courtesy of Richard Bush

Holley as Gloria in *Handful of Salt Water* at the Neighborhood Group Theater in New York City in the 1980s. Fellow actor David Wilber kisses her shoulder.

would not be a good long-term fit. She promised herself she would find something else to pursue the day she got out of school.

Holley kept that promise. The day after she graduated from dental hygiene school, she talked with a career counselor named Dr. Laff. Holley and Dr. Laff had a long discussion about Holley's interests, and ultimately Dr. Laff suggested that she try anthropology. Holley called the anthropology department at Florida Atlantic University; it was reluctant to accept her as a graduate student until it learned about her experience and knowledge of teeth. She began her anthropology career as part of a forensics project on tooth erosion.

Holley hypothesized that handedness, a tendency to use one hand over the other, would influence which teeth would erode the most, and to prove this hypothesis, she worked with her hygiene school to gather relevant data. When she began to review the data, Holley quickly saw that her hypothesis did not hold up. Discouraged, Holley wanted to get away to clear her senses. She signed up to work on an archaeological survey in the Gila Wilderness in New Mexico hoping for an adventure. She loved being outdoors hiking and looking for artifacts. The group needed someone to make maps, and Holley jumped in to learn mapping skills. She learned that she liked making maps and studying artifacts. When she returned to school, she switched her focus to archaeology.

Because archaeology was new to her, Holley relied on the advice of her archaeology professor, which turned out to be a good strategy. First, he told her to take a class in GIS. Here, Holley was out of her comfort zone since she had rarely turned on a computer and did not think of herself as being technical. She found GIS difficult but spent as much time as she could with the teaching assistant, even hiring her as a tutor. Her professor wanted Holley to go to field school, and remembering her experience in the Maya forest, she went to one in Belize in Central America.

The field school was run by Belizean archaeologist Dr. Jaime Awe, who would become one of the most influential people in Holley's life. Jaime had just begun a project in a cave called Actun Tunichil Muknal, a wet cave running through a long tunnel system located in a remote jungle. The Main Chamber, where the Maya artifacts were discovered, was about 1.5 km from the entrance. "To get to this site, we had to hike over an hour on jungle trails, climb huge boulders at the cave entrance, negotiate the wet tunnel, and eventually crawl into the Main Chamber. With 15 field school students to navigate the tunnel and crawls, this left only about two hours of work time every day. So there was little time for learning, but I did get to see the cave, and I took lots of photographs of artifacts."

The entrance to Actun Tunichil Muknal.

Taken by Holley.

Holley always loved caves.

When I was little, my parents took me to Mammoth Cave [Kentucky] and Carlsbad Caverns [New Mexico], and after that, whenever we passed a sign on the highway for a cave tour, I always wanted to stop. In school I started drawing pictures of caves hoping to persuade them to take me to more caves but to little avail. I had not forgotten these early experiences, so when I found that I could go to field school, visit a cave, and study the ancient Maya, it seemed just too good to be true.

Holley drew this cave when she was seven years old.

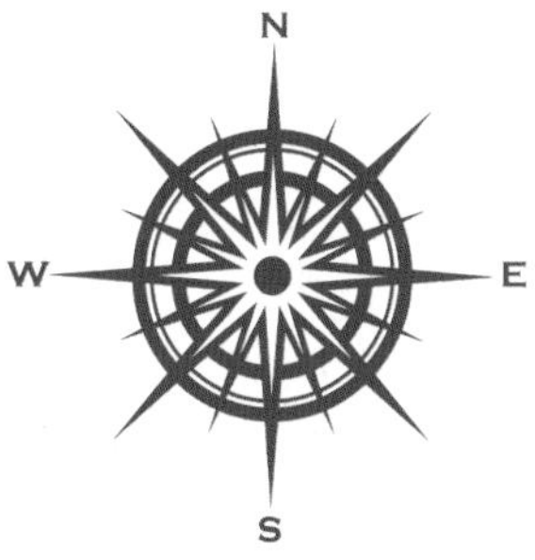

But it wasn't easy. Hiking to the Main Chamber and back each day left her legs feeling like putty. Exhausted, she kept at it but started to lose enthusiasm. When she started working on house mounds, her enthusiasm surged again. Household archaeology might be her niche, she thought.

After returning from the field, Holley wrote a paper (her professor gave her an A+) that she sent to Jaime, who asked her to work on the main chamber for her master's thesis. But Holley wanted to continue working on households, recalling her experience from the summer, and applied for a funding grant. She applied, but she was a new graduate student in competition with more established scholars and did not get the funding. The cave was looking better, and she told Jaime she'd participate.

What she didn't realize was that this cave was one of the most important finds in cave archaeology, not only because it contained intact human remains and thousands of artifacts, but also because it had not been looted. Most cave sites have had a great deal of looting, which destroys the archaeological context and limits what can be done with the material or what can be gleaned at the site. "If I had known this, I would have been terrified, but I was blissfully unaware of the site's import," Holley says.

Because Holley had taken a GIS course, she planned to do a spatial analysis on the placement of artifacts in caves. Once she got into it, she realized that cave archaeology was a good fit for her and continued to work on caves for her PhD at the University at Buffalo, New York. She received a National Science Foundation-funded Integrative Graduate Education and Research Traineeship (IGERT) grant, an interdisciplinary program covering GIS and archaeology. Additionally, she took a track in philosophy and began to work with cognitive scientists and an environmental psychologist named Daniel Montello. She and Dan were interested in cave spaces and how people think about and navigate them. This topic became a growing interest for Holley later in her career.

Holley in Actun Tunichil Muknal while working on her dissertation.

Though it took her a while to get there and it was not her intention, becoming an archaeologist was not entirely an accident or a surprising turn of events. During Holley's childhood, her house in Florida was full of *National Geographic* magazines where she perused excavation photos and read about archaeologists and paleontologists. When she was seven years old, she was inspired to bury her plastic dinosaurs in the sand pile in the backyard and invite her friends over to make discoveries. In this game she created, she was the excavation supervisor who showed her friends where to dig. This game made her extremely popular among the first- and second-graders.

Holley has remained friends and colleagues with Jaime for over 20 years, collaborating on numerous cave projects. Holley helped develop methodology for archaeological mapping based on techniques used by cavers. Using her data, they proposed that ancient Maya religious practice changed during the Late Classic period (AD 700–900), a time of stress brought on by a regional drought. The drought had catastrophic effects, and, as a result, the Maya kingships and political systems failed. People used caves more frequently, and more ritual caves came into use. "We called this the Late Classic Drought Cult," Holley says. "We suspect that this was an attempt to curtail stress and reduce warfare. Most recently I started working on a small Maya city that has a cave system running beneath it. I think that this was an ancient pilgrimage site established as another ritual response to the drought."

Based on her research on ancient Maya caves, Holley began to look at these sites from a cross-cultural perspective. She found that people have lived only rarely in the dark zones of caves but have used them as ritual spaces.

We now think that even Neanderthals conducted special activities in the backs of caves that may have been sacred in nature. This begs the question, Why caves? My new research is partnering with cognitive scientists and psychologists to examine this question, looking at the effects of darkness on people. This inquiry goes to the heart of what it means to be human and is implicated in the origins of supernatural beliefs.

Holley works on a map in the darkness of Ofrenda Cave.

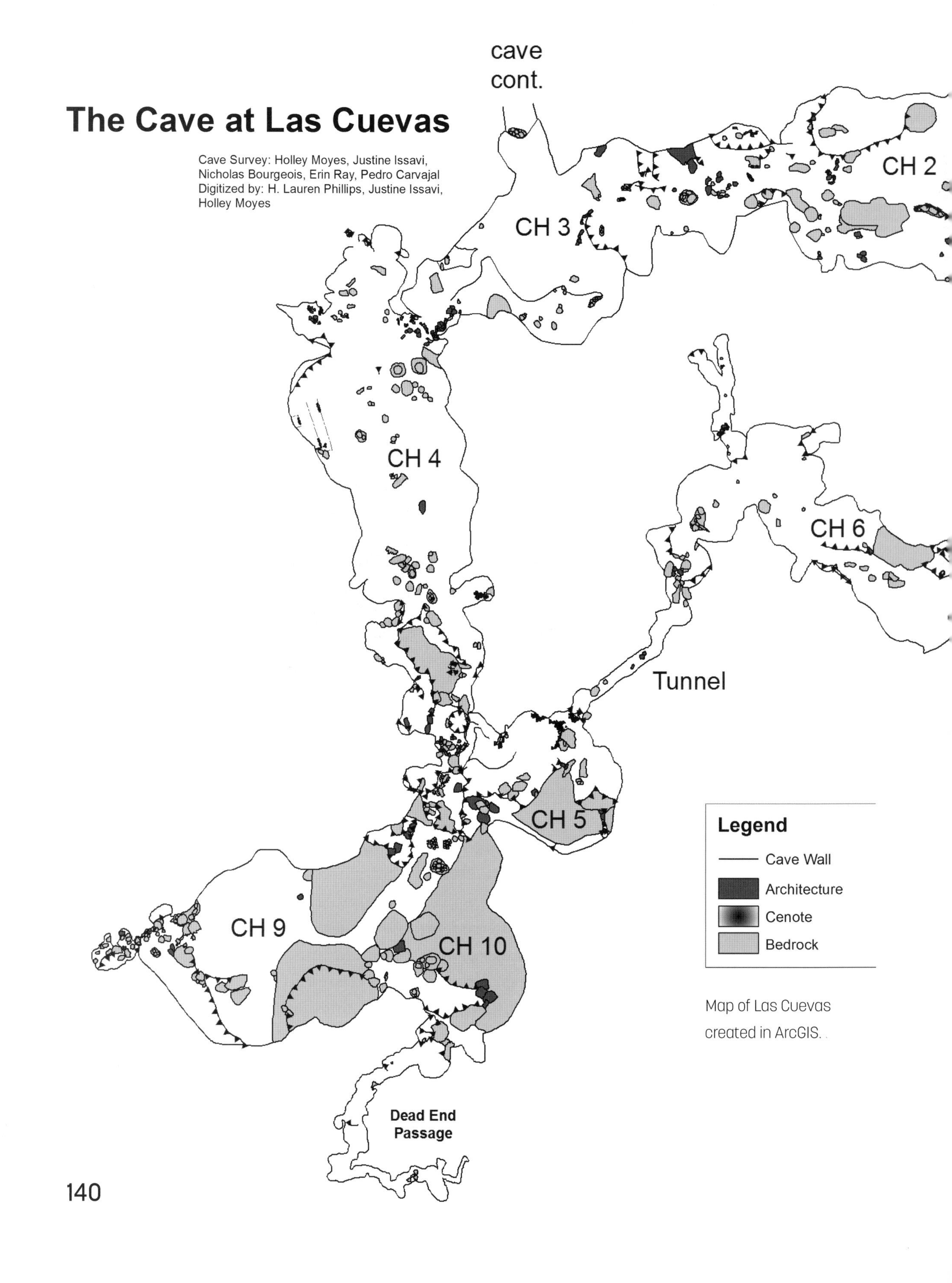

Map of Las Cuevas created in ArcGIS.

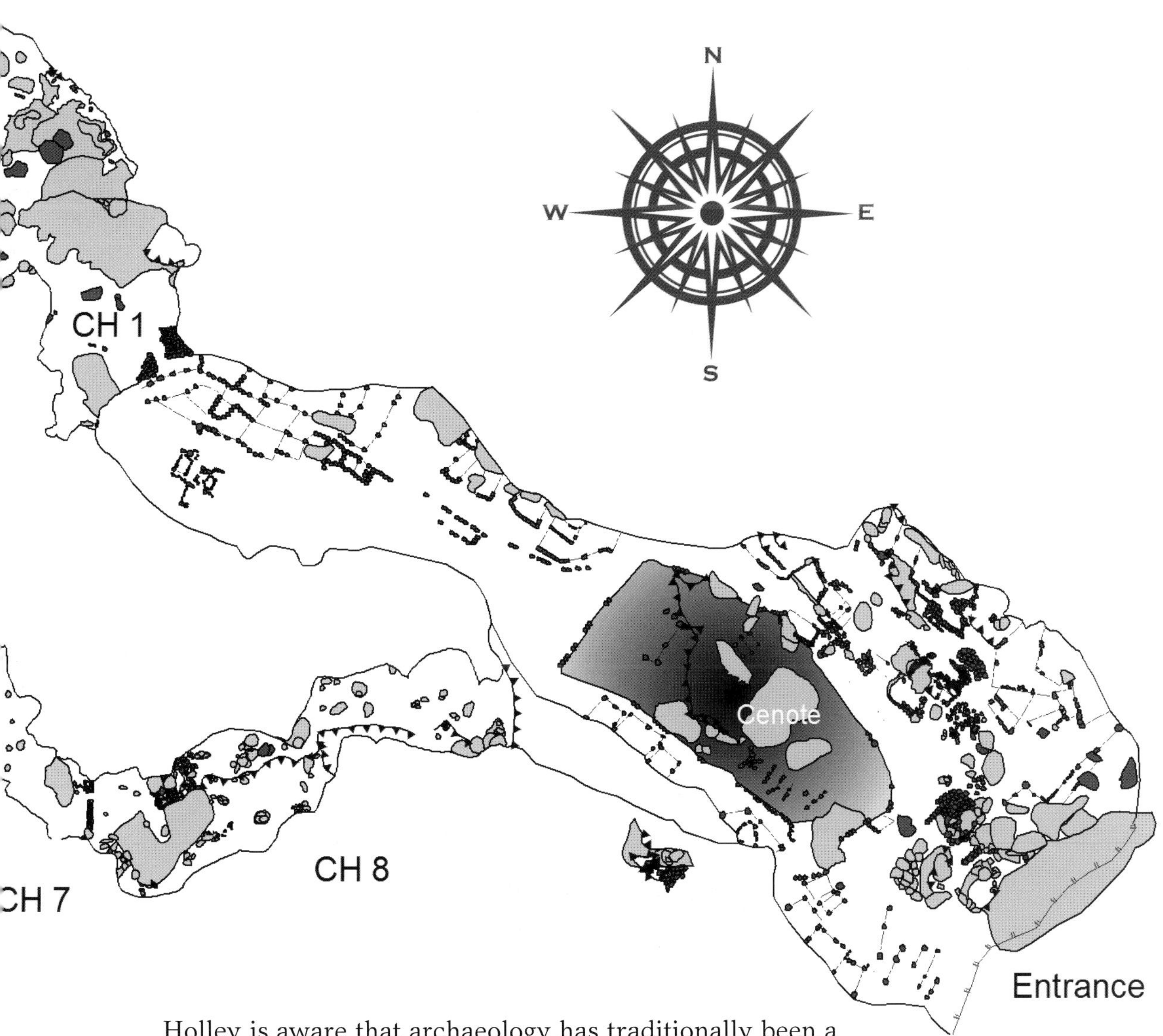

Holley is aware that archaeology has traditionally been a male-dominated field and, to a certain extent, still is. However, a growing number of women are leading projects and working in STEM-related subfields such as bioarchaeology. "I know a number of women doing laboratory research with isotopes and material science. I encourage my students to acquire expertise not just in archaeological theory and practice, but in technologies as well. I think among my most successful students are those who develop good technological skill sets. We currently produce more archaeology PhDs than there are positions in universities. For this reason, I make sure that my students (male and female) are flexible and can fit positions in many markets."

Holley displays the original, large-scale Actun Tunichil Muknal map.

In addition to the growing number of women in STEM fields, Holley believes other changes, such as gender equality, are attainable.

The changes from my mother's generation to mine have been substantial. My advice to women is to never let inequities hold you back. Strive for excellence, and never give up your dreams. Be patient. Things are changing and will continue to do so, but change is incremental, and people need time to adjust. Change must come first from within ourselves. If we want equality, we have to practice equality with everyone.

Be courageous, and don't underestimate your worth. Most importantly, women need to help each other. I have witnessed situations in which women have not aided or promoted one another due to insecurities and jealousies. Sometimes women feel threatened by other women, and I think this is coming from an experience of the world as having limited good. My advice is to remember that you can make opportunities, so work against these feelings. If you can lend a hand to help another woman rise, make the effort. *

Natalia Ocampo-Peñuela

On the wings of conservation ecology

Opposite: Natalia uses birds to further conservation priorities in threatened locales.

NATALIA OCAMPO-PEÑUELA WAS A FIRST-YEAR ECOLOGY STUDENT WHEN her mother inadvertently signed her up for an ornithology course at a nature reserve called Rio Blanco in the Central Andes of Colombia. She didn't know anything about birds, and with her toy binoculars, she could barely see them, let alone study them. One day, she was learning how to capture birds in mist nets when she was allowed to hold and release a tourmaline sunangel, a glittering green hummingbird with a shiny pink throat and a sky-blue spot on its forehead. She fell immediately in love with the bird and at once furthered her passion for tropical conservation ecology. Now Natalia uses birds, and their distributions, to study the impacts of human activities on ecosystems and improve conservation priorities in the world's most biodiverse and threatened places.

Currently, she is a postdoctoral fellow at ETH Zurich (part of the Swiss Federal Institute of Technology). Her research looks at the impacts of oil palm plantations on biodiversity, in the hopes of finding wildlife-friendly landscape designs. In Borneo, she studies the impacts of forest loss and oil palm expansion on the habitat connectivity for seed disperser birds and mammals. In Colombia, she investigates the co-occurrence of oil palms with threatened biodiversity and currently does fieldwork to identify the best forest configuration to maintain biodiversity in oil palm landscapes.

sherpa

Her love of nature and passion for conservation stemmed more from life outside school than in the classroom. She was born and raised in a small town in Colombia called Villavicencio, where the foothills of the Eastern Andes give way to the vast savannas of the Llanos. She lived on farms as a child, and real-life animals were her favorite "toys." As a Colombian native, she couldn't help but fall in love with the megadiverse landscapes and species.

Born in Colombia, Natalia has always loved the natural diversity of her homeland.

Both of her parents were involved in academic research and conservation, so they often took her on field trips and she learned early about the importance of nature. She didn't have a traditional upbringing—she often traveled to exotic places with her parents, returning to school with fantastic tales of seeing and holding wild animals, being attacked by wasps, or playing with indigenous children. Her parents always encouraged her to learn more and share her knowledge. She was never told that she couldn't be what she dreamed of becoming. She grew up thinking she could do anything if she worked hard and respected others.

Natalia studied ecology at the Pontificia Universidad Javeriana in Bogotá, Colombia. During her undergraduate studies, she cofounded and led the first student birdwatching group, Andigena. The group studied the impact of the campus's windows on bird deaths and made a field guide to the birds of the university campus. With the support and help of her undergraduate thesis adviser, Andres Etter, she published research on landscape ecology and birds in the Llanos, which became her first major scientific paper in the journal *Neotropical Ornithology*. Her undergraduate adviser was the first person to teach her the importance of publishing scientific research and striving for opportunities that she felt were outside her reach, such as international fellowships.

Upon finishing her ecology undergraduate degree, she applied and succeeded in getting a Colciencias-Fulbright Scholarship to pursue a PhD. The scholarship, cofunded by the governments of Colombia and the United States, allowed her to do her doctoral studies at Duke University in Durham, North Carolina, with conservation biologist Stuart Pimm. Her PhD adviser became her academic mentor, and they both enjoyed studying birds and applying the results of their research to conservation efforts. This resulted in a PhD thesis that was applied many times, which resulted in concrete conservation outcomes. And it was books by Pulitzer Prize winner Edward O. Wilson, specifically his *Letters to a Young Scientist* (Liveright Publishing, 2013), that got her through her PhD exams by encouraging her to follow her passion.

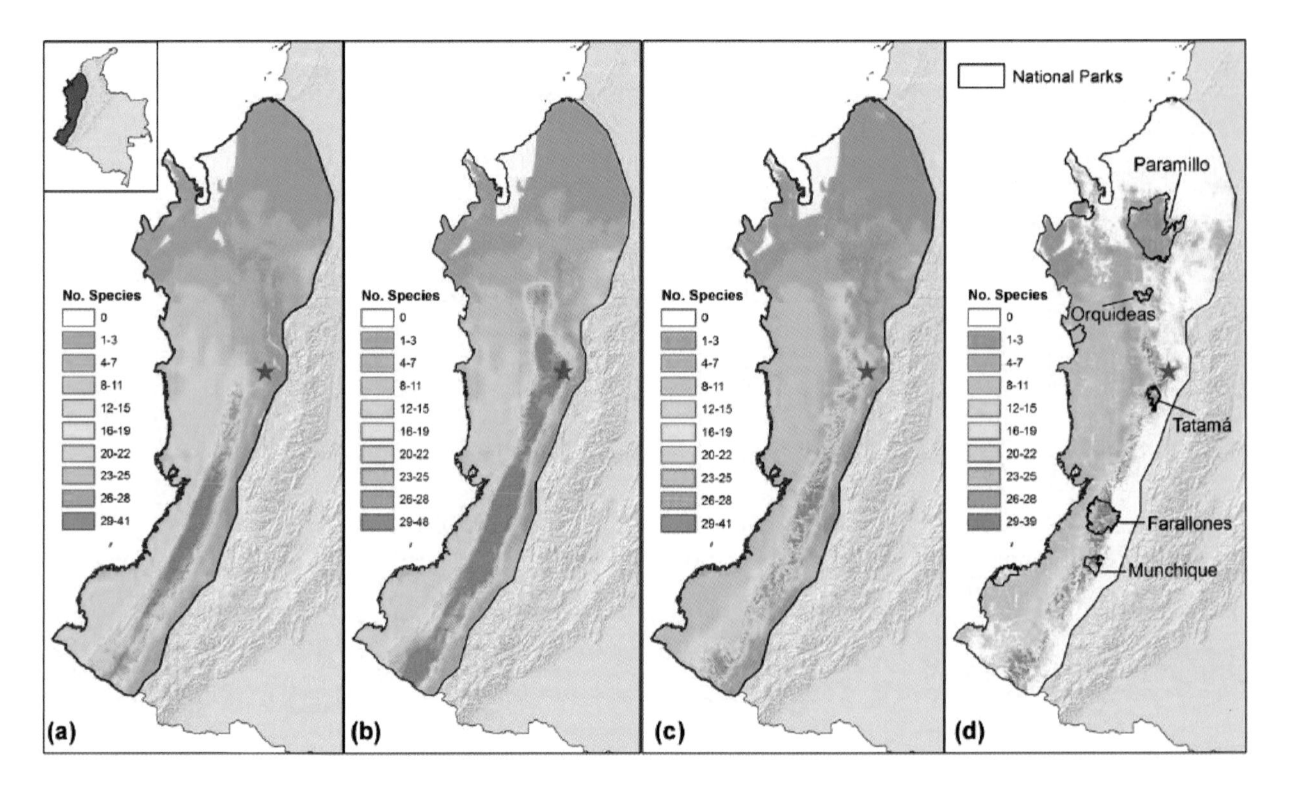

Concentration of 100 endemic and small-range bird species in the Western Andes of Colombia, excerpted from Natalia's first published research paper.

During her time at Duke, Natalia published four papers about using bird distributions to improve evaluations of extinction risks and better set conservation priorities in Colombia and five other biodiversity hot spots. Her first objective was to find the most cost-effective place for conservation funding to restore forests in Colombia. Using GIS for the first time, she mapped distributions of endemic and range-restricted birds using publicly available data and information gathered during her own fieldwork. The Western Andes forests near Jardin, Antioquia, came up as the biggest conservation priority because of their high concentrations of endemic and threatened birds. Natalia's research led to funding to restore a forest that would reconnect large forest fragments and benefit at least 45 endemic and threatened birds.

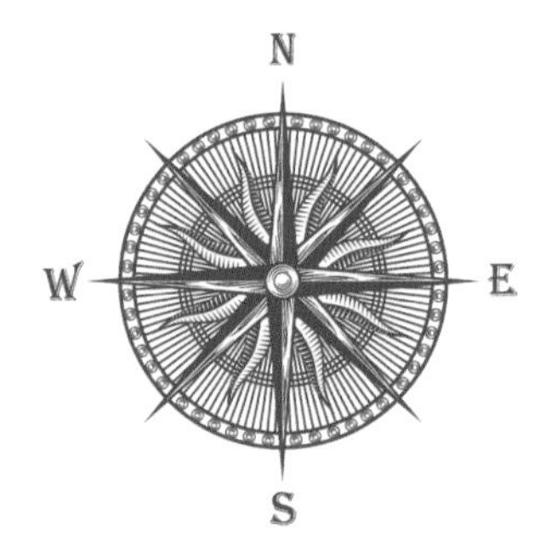

Natalia's research of birds has led to the protection of many threatened species.

Natalia's capstone paper dealt with the need to update the priority "red list" of endangered birds used by conservation groups on the basis of the amount of remnant habitat for species already deemed threatened, and for those erroneously classified as nonthreatened. Through her research, Natalia found that many more birds were endangered than previously thought. Studying 600 species in six biodiversity hot spots, she demonstrated that using remotely sensed data of elevation and forest habitat could greatly improve extinction risk assessments for forest birds. This research resulted in an agreement with the International Union for the Conservation of Nature, and she is on target to publish a joint paper recommending that species ranges be refined by elevation and habitat when practical.

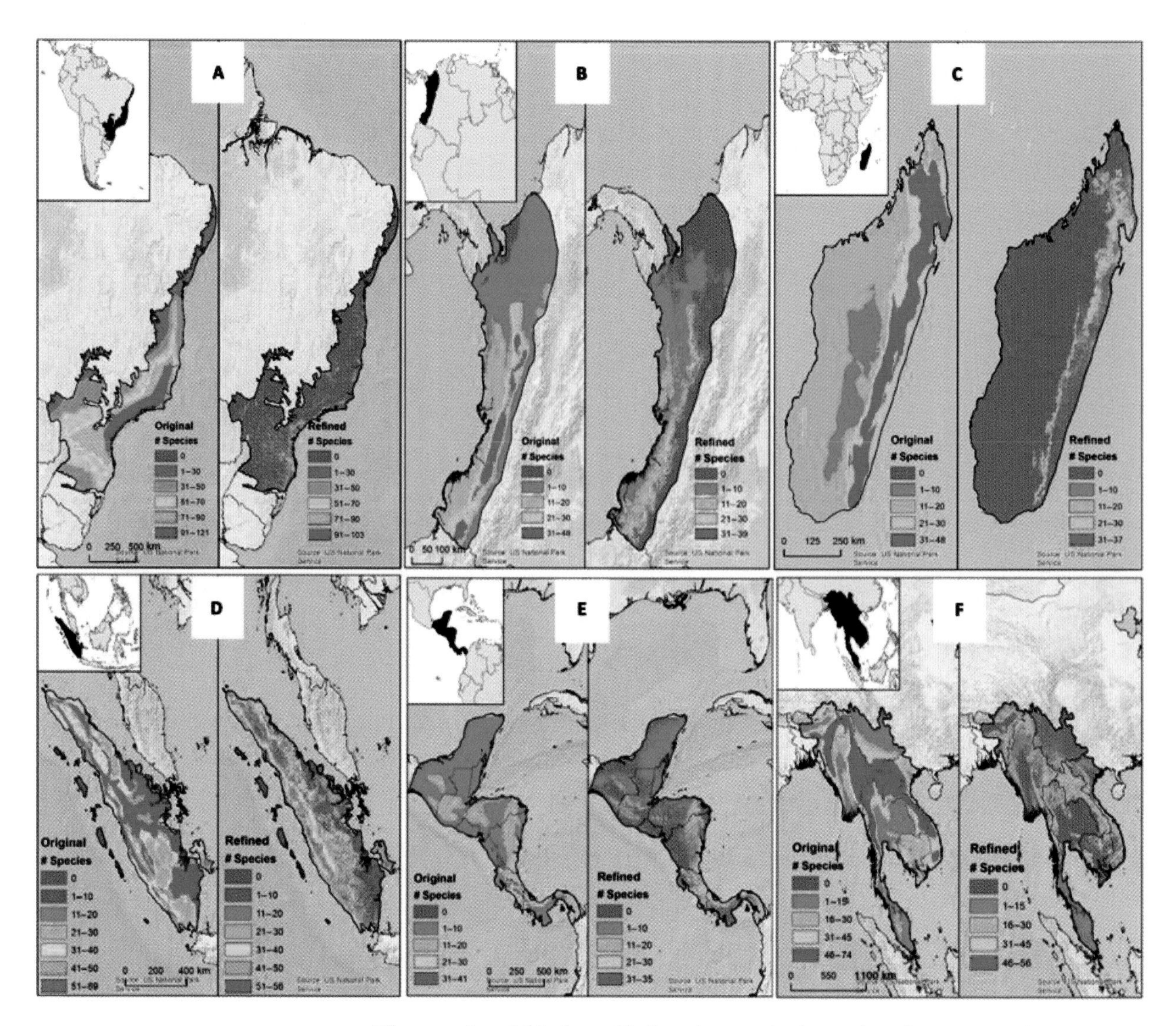

Concentrations of endemic and threatened bird species in six biodiversity hot spots: the Atlantic Forest in Brazil, Western Andes in Colombia, Madagascar, Sumatra, Central America, and Southeast Asia.

The study of birds colliding into windows has been a "side project" of Natalia's since she was an undergraduate. As a PhD student, she started and led an effort to document and prevent birds colliding into windows on Duke's campus, which led to some major achievements. She and her collaborators were able to document and map these bird collisions into windows of several buildings around campus during peak spring and fall bird migrations, using the app iNaturalist and GIS to identify areas where collisions were more prevalent. This study provided the motivation for the university to retrofit the deadliest of buildings to make the campus more bird friendly. She sees this local impact as one of her biggest and most tangible conservation successes.

Part of Natalia's work goes to restoring her beloved forests.

Of course, life always has its obstacles, even in academic careers. While living in Colombia in a tent on a forested mountain for a few months, she came to find out that her salary for doing fieldwork had been cut off. It was eventually resolved, but similar difficulties underscored the constant struggle to have academic work appreciated monetarily. During the same campaign, she woke up with a terrible toothache. The first inexperienced dentist she went to managed to drill right through her tooth, so she decided to make the trek to the closest city four hours away to find a dentist. The verdict: the small-town dentist had damaged her tooth permanently, and it had to be pulled out immediately. Crying in the dentist's chair and wearing her muddy boots and dirty field clothes, it didn't help that when they asked where she lived, she had to say, "Inside a tent on a forested mountain." Undoubtedly, the dentist must have concluded that she belonged to one of the many armed guerrilla factions that have plagued Colombia's national security for decades.

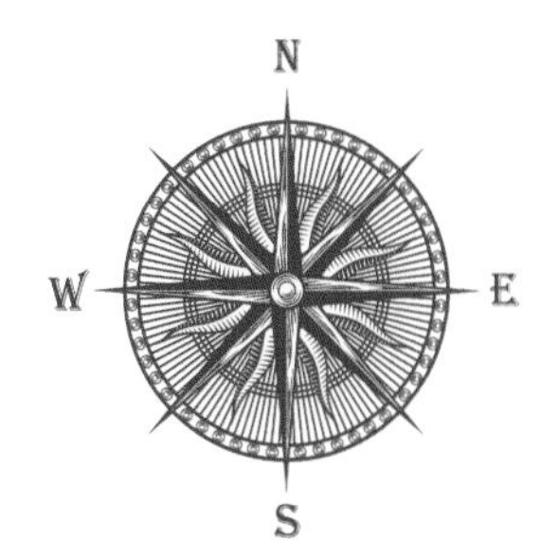

As a Latina woman, Natalia understands that hurtful comments and attitudes, whether race- or gender-based, can influence self-esteem, and she encourages women in general to ignore these comments and continue pursuing their dreams and doing good research. She has often suffered from her desire for perfectionism, but when planning fieldwork in tropical nations, "perfect" is far from reality. Many nights spent in tents in the forest and on remote islands have taught her that being adaptable is better than being perfect. She has the same philosophy when it comes to conservation.

We can't have it all, but we can create sound priorities that allow us to protect the most vulnerable species while we work towards securing more funding for wider protection.

Currently, she is facing the dual challenge of balancing life and work, of having a family while pursuing an academic career. She sees that even as women are encouraged to go into STEM fields, they tend to leak out of the academic pipeline because of the difficulties in obtaining a professorship. Recently, she has been faced with her own hard choices. Having finished a postdoctoral position and with prospects of another, she and her partner have started a family. Given the current statistics, she hopes that this decision will not inevitably hurt her career. She feels that society has a long way to go to achieve equality but sees hope in cases where maternity leave is improved, women are evaluated in an equitable manner, and flexibility is offered for balancing work and home life.

Natalia, *right*, with her field team in their forest camp.

Ultimately, Natalia wants to influence conservation decisions in Colombia and elsewhere. Having her research be noticed by the government, NGOs, and other conservation advocates is the key to successful conservation, she says. She is hopeful that conservation ecology is gaining momentum and that there is increasing awareness about the loss of biodiversity and of climate change and pollution. She sees the interdisciplinary focus of conservation ecology and its reliance on increasingly innovative technologies, including drones, apps, remote sensors, tracking devices, and recording devices, as the future. Making the best use of these technologies will help combat wildlife crime such as hunting, poaching, and the pet trade, she says. And it will monitor animal movements and their response to environmental change as well as help engage the public in conservation efforts.

Even more importantly, she hopes her research inspires other young scientists, especially Colombian scientists, to pursue their passion, study biodiversity, and work to help protect it. She would like her work to influence future generations, starting with children. Some of the most rewarding times of her life have been spent in a small room at a rural school teaching about birds, she says. Natalia hopes that her passion for her country and its birds will serve as a role model to others to be better scientists, accomplished conservationists, and good mentors in the pursuit of their dreams and causes. ✸

Natalia looks toward a horizon of improved conservation.

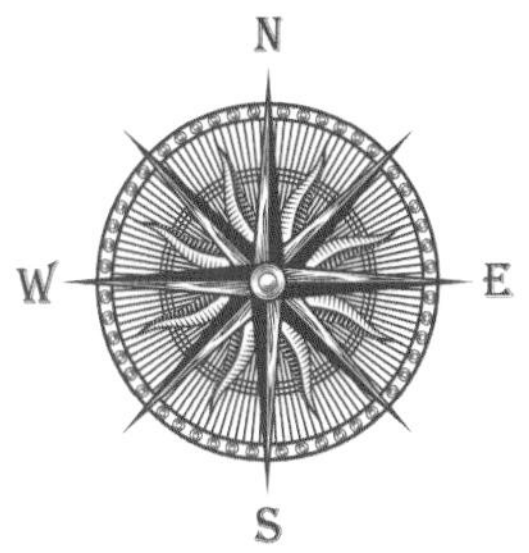

Miriam Olivares

Mentoring from Mexico to Yale

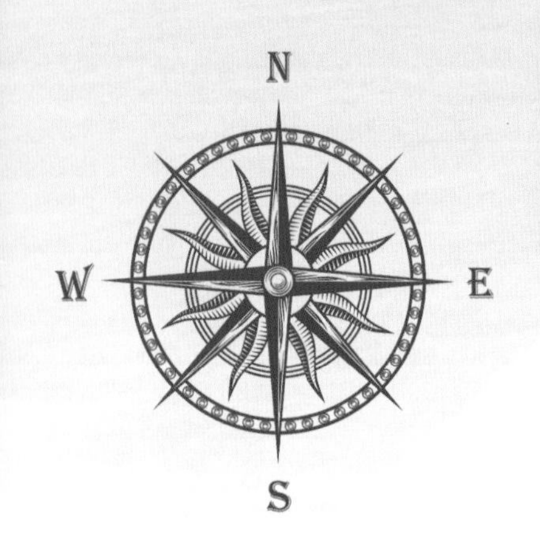

MIRIAM OLIVARES DID NOT GROW UP READING BIOLOGY TEXTBOOKS OR collaborating on science projects. She spent her childhood surrounded by the flora and fauna of magnificent landscapes—her backyard full of waterfalls and birds. The food, the colors, and the people of Mexico were her laboratory, fostering her love of nature, her empathy for the poor, and her celebration of life.

This love of nature, along with hard work, leadership, creativity, and perseverance, has propelled her to her present position as the GIS librarian at Yale University. She coordinates the GIS services at Yale, including revamping and establishing new GIS services, teaching GIS workshops, sharing expertise on GIS apps across disciplines, and advising or connecting experts across campus, regionally and globally.

Miriam, the fourth of five children, grew up in Ciudad Valles, a small city located in the Huasteca Potosina region, in the northeast of Mexico. As a child, she spent most of the summers at her grandparents' farm and, later in life, she constantly visited her parents' farm, Ojo Caliente. Growing up in Mexico taught her to be strong and resilient, and instilled in her the importance of creativity: less is more. When she hears, "We have no budget," what it means to her is, "Find a creative way to do what you want because we cannot help you with funding."

From kindergarten to college, nobody talked to her about aiming big. She does not remember teachers encouraging her to study science. However, she fondly remembers how her parents worked hard to provide her an education. Her father was the first student in his town to graduate from college, while her mother was taken out of elementary school to take care of her grandmother. Miriam is proud of both parents: she always admired her mother's selflessness and her engineer father's ability to build "empires"—her perception as a child—without many resources.

Education was always important to Miriam's family; pictured, Miriam, *second from right*, in her fourth-grade class.

Her older siblings were her real inspiration. She remembers being fascinated by the intelligence of her brother, who could answer any of her questions and went on to get an engineering degree. Ciudad Valles did not have a high school or a university. With her father's blessing, Miriam moved to Monterrey, Mexico, at the age of 14 to join her siblings. She studied in "La Prepita" (Monterrey Tech High School), which she found amazing but rigorous, difficult, and challenging. It was a very different world, at least for a girl from La Huasteca.

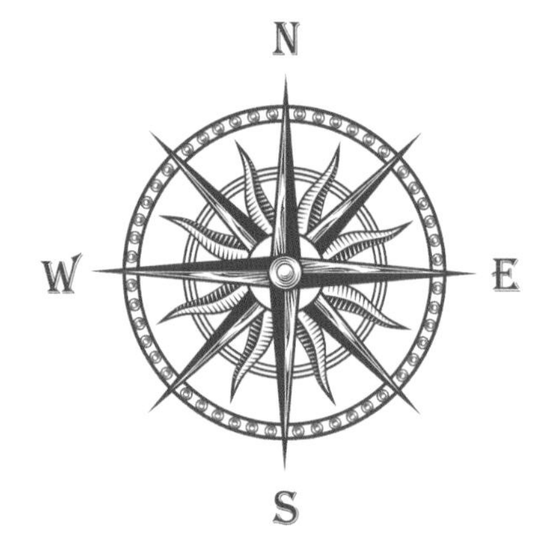

When she completed high school, all her siblings had finished college and left Monterrey. Consequently, her conservative father wanted her to return home—he would not allow his young daughter to live alone in a big city. Miriam now says she knows that problem would not have occurred if she had been a boy. Her Monterrey-native high school friend Rocio learned of her problem and invited Miriam to live at her family home, obtaining permission for Miriam to continue on to college. Looking back, neither of them realized then the tremendous impact this act of kindness would have on Miriam's future.

In college, she was unable to narrow her concentration; she enjoyed many fields of study—engineering, science, mathematics, and psychology. She ended up choosing architecture because of its interdisciplinary nature. She was involved in musical theater and student leadership. She participated in a fieldwork internship and summer design program. Monterrey Tech had instilled in her a strong foundation of leadership, innovation, and problem solving. She graduated as an architect, went back to Ciudad Valles, and opened her own architectural firm and business. She was only 20 years old.

Years later, Miriam did a 180-degree turn, but her decision to move to the United States was not planned. By then, she was a recent divorcee with a three-year-old son. After visiting a college classmate and his family in College Station, Texas, she decided to stay to practice and improve her English, and so enrolled for ESL classes at Texas A&M University. Looking to learn technical vocabulary, she sought to audit a class in the College of Architecture; she was rejected 10 times over the course of a year because of her "limited English," because, she was told, she "would not understand anything." Finally, Professor John Alexander allowed her to audit his course on History of Architecture. From that experience, she learned that to get what you want in life, you must ask at least 11 times.

She applied and got admitted to Texas A&M, where she studied for a master of science in land development. While there, she went to study abroad in Australia, where she was exposed to GIS for the first time. She finds invaluable the support she received from her mentors, Dr. Rick Giardino and many others, all of whom showed kindness to her and her son, Ricky. She was offered a full-time job while conducting her doctoral work, specifically to build the GIS services at Texas A&M University Libraries, and then at the school's Center for Geospatial Sciences, Applications, and Technology (GEOSAT). During her tenure, she was the backbone of GIS services at Texas A&M, helping support the entire university while researching her own topics.

Miriam at Texas A&M University Library on GIS Day 2011.

Among other projects, she took part in multiple marine science expeditions to predict and verify areas where reef fish aggregate to spawn in Belize, Mexico, and Guatemala. Her team, led by Dr. Will Heyman, put an emphasis on cooperative research with local fishermen. This research allowed Miriam to return to her roots and contribute to sustainable management at home. She loved the opportunity to work with local communities to share a mix of high-tech science with local ecological knowledge to address societal issues. The data generated has been used by Miriam and others to map previously undescribed coral reefs and establish several marine protected areas. At A&M, Miriam worked with professors and many GIS peers who honed her learning. She received her GISP certification with the GIS Certification Institute, all while being a single mom on her own in the United States.

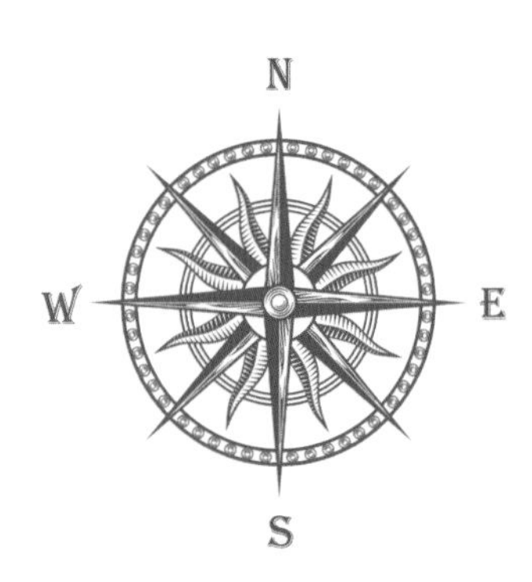

Miriam, in red top, in Quintana Roo, Mexico, with fishermen and scientists from her Mexican partner organization Community and Biodiversity (COBI in Spanish).

Miriam never saw motherhood as an obstacle to work. Her son Ricky was both her main source of strength and her biggest challenge because she devoted herself to providing him a better life. Everything else could fail, but regardless of the hardships, she was determined to give Ricky a sense of peace and stability. She often felt guilty about the long working hours but appreciates that this reality helped her son become independent and resilient. She also had a group of wonderful Latin "aunties" who adopted her son as their nephew and did everything they could to help give him a "normal" family life.

Currently at Yale, Miriam coordinates the GIS research support services across campus, working to establish collaborations on campus and beyond. She organizes GIS Day on campus and strives to

Miriam and her son, Ricky, in 2006.

expand the reach of GIS across academic departments. For Miriam, the gift of knowledge has always provided something bigger in return.

Miriam considers herself a GIS evangelist who loves talking to students from underprivileged communities. She advises them to stay alert to when the opportunity arises.

Say no to fear, work hard, and the reward will come.

Saying she wishes she had believed in herself earlier, Miriam also recognizes the effect of having great mentors and the importance of having systems in place to ask for help. A believer in the ability of GIS to change students' destiny, Miriam plans to continue and expand her commitment to education by mentoring students who need a little push to get into STEM and the rewards that await.

GIS has opened opportunities for her to address challenges, directly through her own research and indirectly by enabling other researchers to do their work. Miriam plans to continue guiding and educating those who have the power to change our world by helping to meet their research agenda. She believes that GIS educators are in the business to enable others to better our world. And she plans to facilitate that mission by continuing to share her passion for GIS and geospatial technology with those seeking to make a difference. ✸

Breece Robertson

Protecting what she holds dear

BREECE ROBERTSON WAS ALWAYS CURIOUS ABOUT HOW THINGS WORKED, especially in the natural world. Growing up in Hamlet, a small town in North Carolina, she reveled in the green woods, searching for snakes, building forts, fishing in creeks.

Being so intimate with nature and the outdoors sparked my love of science.

Breece wondered why animals behaved the way they did and tried to understand the stars in the night sky. She wanted to learn, and science was her path. From learning about combining the elements in chemistry class in sixth grade to studying how the body works in physiology classes during her undergrad years to understanding how the earth's wind and ocean currents work in concert in earth sciences courses during her master's studies, she has always been motivated and enthused by science-based subjects.

I have a deep appetite for learning and have always consumed as much knowledge as possible to further my understanding of how our planet and universe operate.

N

W E

S

Breece continues to learn about the natural world as she practices conservation by creating parks and protecting natural lands, ensuring healthy, livable communities for generations to come. A vice president of The Trust for Public Land (TPL), Breece runs the research and innovation department, consisting of 18 staff and 10 or more consultants nationwide who use GIS for planning and research to strategize TPL's conservation approach. Breece skillfully uses GIS maps, data, and analysis to communicate the trust's mission of protecting public lands to colleagues, policy makers, partners, and affected communities.

In grade school, Jane Goodall was one of Breece's first inspirations, and she has been following her achievements ever since. Breece appreciated Jane's studies of primates and her desire to change women's roles and advance women in science.

I have always been drawn to strong women who aspire to learn and move the field forward while paving the way for other women to do the same.

Even to this day, when faced with a dilemma, Breece will ask herself: What would Jane do? She was also inspired by, and followed, great explorers who focused on science such as Charles Darwin and Alexander von Humboldt. Sylvia Earle, featured in this book, is another influence: "I love that she is on the road more than 325 days a year educating about the importance of the oceans and the wonders that lie there that we must protect."

At first, Breece thought she might be a veterinarian because of her love for her dogs and her interest in animals. She excelled in science and attended Lenoir-Rhyne University in North Carolina to study exercise physiology and math. After getting her undergraduate degree, she thought she might become a doctor. Knowing that she would have many years of school and residency ahead, she decided to take a break. She had never been west of the Mississippi River, so she packed a suitcase and loaded her dog Maggie into a van and headed to the Intermountain West, camping and exploring throughout this new landscape. She fell in love with the western outdoors and lived in Albuquerque, New Mexico, for five years. Working as a physical therapist, Breece realized that medical school was not for her. She wanted to do something to protect access to the public lands that she had enjoyed for hiking and recreation. She wanted to make sure other people could enjoy them, too.

Moving back to North Carolina for family reasons, Breece attended Appalachian State University, getting her master's degree in geography and planning. There, she studied a variety of subjects, from feminist and regional geography to GIS and statistics, and collaborated with peers and professors, in preparation for a professional life where collaboration is essential. Her parents have been her biggest supporters, and Breece considers herself fortunate that they backed her education and stood by her while she figured out her options.

Yearning to return to the West, Breece's first job was teaching GIS to US Forest Service and government employees in Boulder, Colorado. GIS had been part of her master's program, but she was initially daunted to teach it to others. "Nothing like being thrown in the fire to really learn something well," she says.

Though she enjoyed teaching, she wanted to do more. She had learned about TPL during her first week on the job when she taught a group from there how to use software for public planning. A year later, she called her contact from that group and said she wanted to work for a mission-focused organization that supported "land for people." She was hired as a consultant, and then soon became the first full-time GIS employee for TPL in 2001. She was able to move to Santa Fe, New Mexico, a city she loves, and knew she had found her dream career because of the passion and commitment shared by her colleagues at TPL.

Breece at the San Juan River near Bluff, Utah, with her dogs, *from foreground*, Cody, Maple, and Ellie.

Breece started her career at TPL creating maps that staff working with landowners could use to tell the story of why a certain property should be protected and what the multiple conservation benefits would be. As she started traveling around the country to TPL's 40-plus offices, she began to see how TPL could help communities develop conservation plans. Working with the Seattle office and her small but growing GIS team in 2005, Breece helped create a process and approach called Greenprinting. Greenprinting brings the voice of the community together with data-driven analytics and maps to chart future conservation efforts. Interactive maps highlight key areas for protection based on residents' conservation priorities and on where possible conservation funds are available. She and her team have completed hundreds of Greenprints to date.

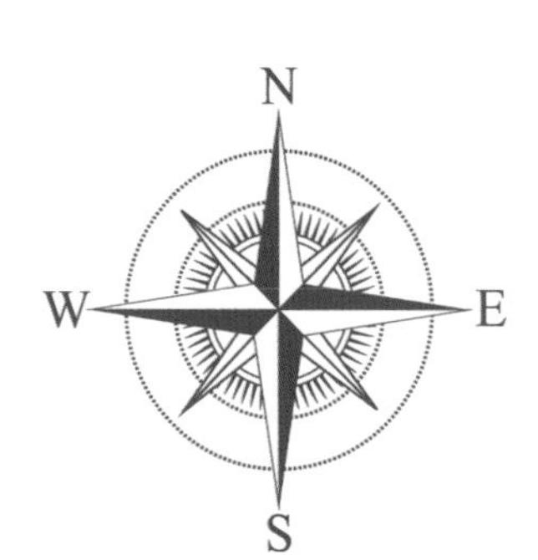

Using state-of-the-art methodologies such as artificial intelligence, deep machine learning, and feature recognition combined with applied analytics and research methods, Breece and her team are working to strategically identify where parks and protected public lands will provide the biggest benefits. Cities and towns throughout the US are using their approaches and data to inform decisions and public policy and direct public money for parks and open spaces where they are most needed.

One of the projects Breece is proudest of is ParkScore®, the first methodology to rank park systems in the largest 100 US cities. ParkScore (www.parkscore.org) uses GIS to map the 10-minute walk

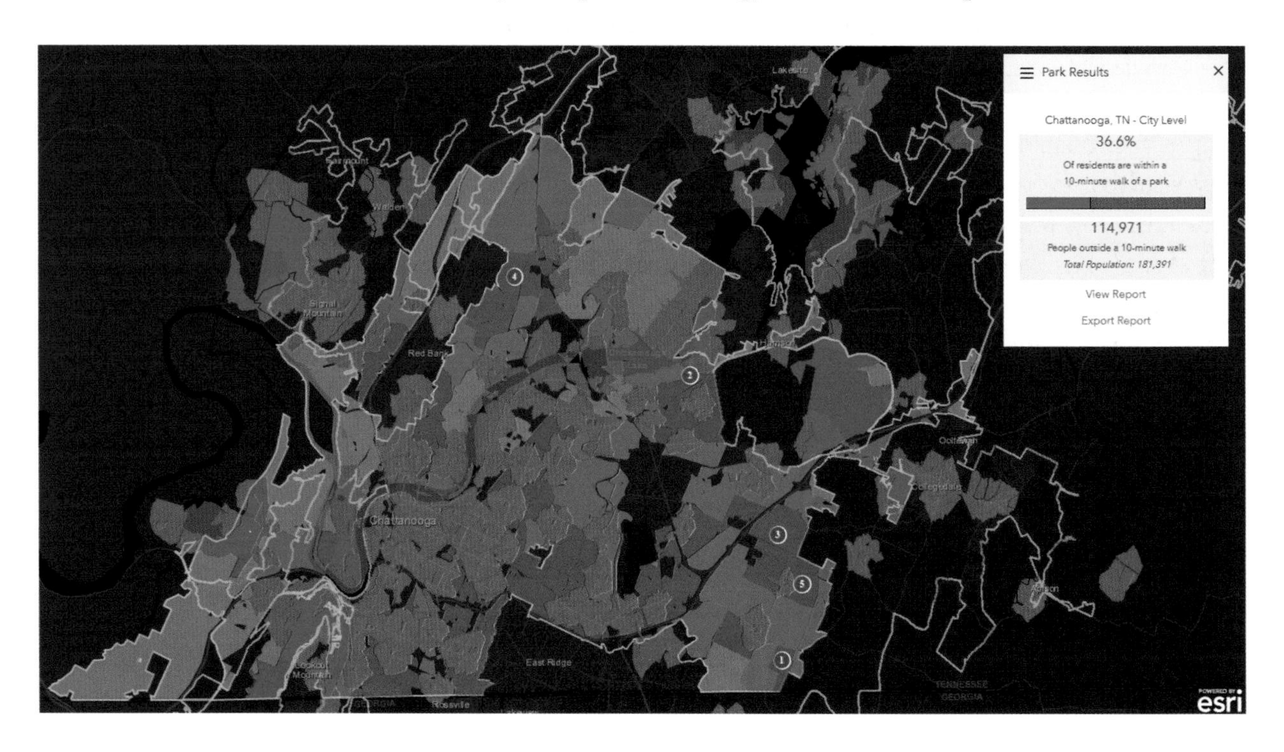

Results from the parkserve .org web app created by The Trust for Public Lands.

to a park using walkable road networks, plus acreage, facility, and investment metrics, and then identifies where people have access to a park and where they don't. TPL and its partners use the maps to see where people need parks the most based on demographics and proximity. She has also worked on ParkServe®, a platform and database of park accessibility for over 14,000 towns and cities nationwide (parkserve.org). Breece's team spent three years contacting these locations for data and creating it where the cities and towns didn't have any. The projects keep growing: next up is Access Impact Mapping (AIM), which focuses on recreation access to public lands (https://dev.tplgis.org/AIM).

Breece and TPL have been awarded twice by Esri for their GIS service in innovation that helps communities meet park and conservation goals. She shares with Esri founder and president Jack Dangermond a passion for conservation.

Breece wants to influence anyone who cares about making the world a better place, both globally and locally, and enjoys collaborating with people who are working to make communities sustainable for people, animals, plants, and natural resources. Breece loves giving keynote speeches and encourages young people in science fields to develop their communication skills. In her job, she has done a lot of hiring and has seen a lot of resumes. Typically, she sees young people emphasize their technical "hard" skills, such as exactly what they studied and what software programs they use. Although that information is important, they often leave out what Breece calls the "soft skills": Do you know how to collaborate with others? How good are your communication skills, both verbal and written? Can you present effectively in front of others? She is convinced that these soft skills are just as important to a person's success as knowing the software.

Breece encourages people to find organizations that embody their passions and to express why they care.

Don't be afraid to ask questions or to state your opinion or hypothesis–even if your voice shakes.

Breece grew into an accomplished speaker, but when she first started, her voice shook and she was nervous. Practicing speaking helped, and she's learned that nervous filler words such as "um" can be transformed into strategic pauses at just the right moment to make an impact. Caring so much about her message has also helped.

For me, thinking about inspiring people to care about parks and open space makes me less nervous because I care about them, too. The more people we have working together to make a difference, the better. And who better to inspire than us?

Breece talks with then-mayor Mitch Landrieu, *far right*, about TPL's Climate-Smart Cities project in New Orleans.

Breece recognizes that other women have paved the way for her and is thrilled to see more women advancing in the sciences faster than ever before. "I just saw a statistic that more women than men are entering the sciences," she says, "though more work needs to be done to make STEM and science-based opportunities available to all women and women of all nationalities, ethnicities, and income status. ... We still face issues with equal pay, equal advancement opportunities, and leadership and executive opportunities." Regardless, she is inspired by what she sees.

We are definitely at a unique time in our history on this planet for women to step into the power that we have to change the world.

Becoming, and being, a leader is something Breece has had to learn in the workplace. Leaders at the top with big visions are inspiring, yet leadership can happen at all levels, she says. She has seen leadership from her peers and from people who report directly to her who are stepping up in different ways. She loves to watch her colleagues lead in both bold and quiet ways.

The world is such a big place, and there are so many things to explore, to care for, to take part in.

She encourages others to "learn from our past but stay on top of the future and the latest innovations, challenges, and opportunities. Never stop learning, and explore as much as you can." ✸

Elena Shevchenko

Forging an international bond in tense times

FOR RUSSIAN BUSINESSWOMAN ELENA SHEVCHENKO, MIDDLE AGE WAS the start of a remarkable second career. Surviving World War II and the breakup of the Soviet Union, Elena helped found the GIS market that would cover over 11 percent of the land on Earth. She is known as "the grandmother of GIS in Russia," yet she is quick to qualify that:

> **I am neither a scientist nor a great specialist in geographic information systems. My mission here was to organize the distribution of GIS technologies in the Soviet Union and then in Russia and in other countries of CIS (the Commonwealth of Independent States).**

After being introduced to Esri® software in 1989, Elena formed a GIS distributorship called DATA+. Now 30 years later, as many as 4,000 organizations use GIS in Russia and CIS, and in a population of more than 150 million people, there are countless users. Elena helped transform a country whose cadastral data was once largely off-limits to one in which maps of 100 percent of the country's land holdings are now publicly available over the internet. Elena's DATA+ helped bring this about by working with the Ministry of

Beginning in the 1990s, with Elena as catalyst, organizations in Russia using GIS went from zero to 4,000 in less than 30 years.

Economic Development of the Russian Federation to consolidate all the data. So how did a 52-year-old woman, with no formal education in business, help integrate geospatial coverage of more than a tenth of Earth's land area? She believed in the promise of technology and the power of human relationships built on trust.

At 82 now, with her business in the hands of her daughter, Yulia Bystrova, and other colleagues, Elena is retired, but her legacy lives on. Lawrie Jordan, who as cofounder and president of the US imagery company ERDAS worked with Elena decades ago, credits her friendliness and diplomacy for successfully introducing geospatial technology in a time of political upheaval. "She believed in doing the right thing, following your gut, and being honest and ethical," Jordan recalls. "She was a good judge of character in handpicking her subdistributors and in selecting the right people to work around her. A thoughtful listener and brilliant conciliator, she was the center of gravity and the moral compass of any group she was in. She helped us work through problems, and there were plenty of those at the time. A remarkable person—everybody loved her."

The staff of women of DATA+ on International Women's Day in the late 1990s. Elena is second from right in the front row. Half the staff she hired were women and half were men.

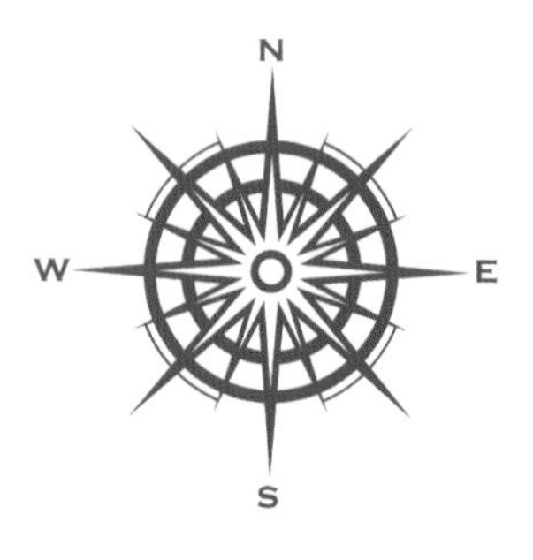

Despite Cold War tensions, Elena helped bridge the divide between two countries that had been hostile for more than 40 years. The new technology was made in the US, and Elena faced problems in getting the GIS software through customs. Each country used a different accounting system and a different language. Her profession as a translator and interpreter within the Soviet Academy of Sciences in Moscow helped her overcome the language barrier, and her work there had other positive benefits as well. By the time GIS first caught her attention, she had become assistant to the director of the academy's Institute of Geography, and because her boss, I. P. Gerasimov, was also chairman of the Environmental Commission of the International Geographical Union (IGU), she attended many of its conferences.

At the 1989 conference in Seattle, Washington, Elena observed a demonstration of PC ARC/INFO® software. She thought it would be wonderful if they could have such a program in the Institute of Geography. Esri then donated PC ARC/INFO to her Institute of Geography at the Soviet Academy of Sciences.

That is wonderful, she thought. There was only one problem: neither the Institute nor any group of geographers locally had the funding necessary to promote GIS in the country. So, in her early 50s, in the face of severe political and logistical challenges, Elena started her own business, the distributorship DATA+, to help promote GIS.

With vast petroleum deposits located within the CIS, hundreds of petroleum companies operate in the region. Elena's efforts to help introduce GIS technology to these regional oil and gas producers came at a critical moment in history that helped augment and preserve the economic ties among newly independent countries.

Today, Elena matter-of-factly recounts the beginnings of an enterprise whose timing brought seemingly insurmountable obstacles. She and her younger partner, Sergei Rostotsky decided to organize a venture which became a distributorship. "Sergei died in October of 1992," she says, "and I was left alone managing the company."

Only seven months into the company's founding, after the breakup of the Soviet Union, the newly drafted businesswoman had lost her business partner.

It all started in one room: Elena's first DATA+ office, circa 1992.

For the new founder of DATA+, however, this was not her first life challenge.

Elena recalls the occupation of her village of Dolinovka by fascist troops in World War II, when she was a young child and her mother was left alone with three small girls when her father went off to war.

"I remember how in the summer of 1943 the German soldiers rushed into our village on motorcycles, and for the first time in my life, I saw grown men in shorts. This made me giggle at first, they looked so funny, but then we saw that there was nothing amusing at all about their arrival: they began running about the village killing cows, sheep, and hens. My mother had hidden a couple of geese in the garret, and when two German soldiers heard the sound of them, they nearly killed her, demanding that she give them the geese. My elder sister, three years older than I, knew enough German to cry out, 'Do not kill my mother—she will give you the geese!' The soldiers were surprised to hear their own language from her mouth, perhaps even impressed, but still my mother had to give them the geese.

"In the middle of winter, the Germans decided to drive the local population away from their homes. This was terrible for our mother—in the cold weather there was no place to go or to hide with small children in tow. Happily, finally, Soviet troops approached the village, prompting the Germans to leave. We heard bombshells flying through our yard, and watched planes fighting in the air.

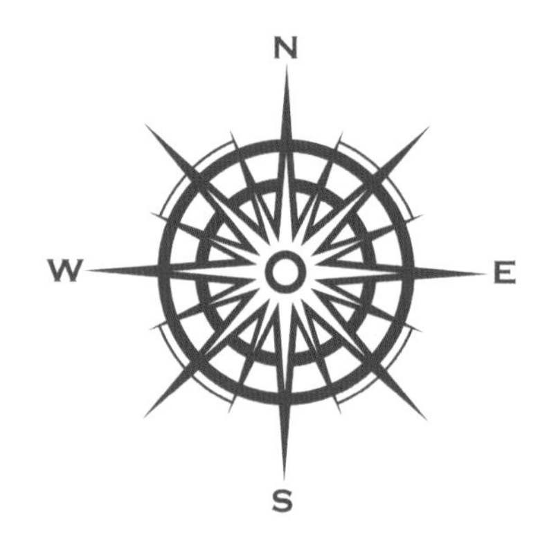

"Our father came home in 1946, and the fight for the life and health of us children began anew. Those were years when most of the produce grown in the villages was taken away from us, in order to support the rebuilding of destroyed cities and industrialization. There was a seven-year school, but not enough textbooks and notebooks. We had to write on newspapers. Nevertheless, I was an excellent pupil, and after completing school, I was sent to a technical college in Krasnodar, where I studied for four years. After graduating, I was sent to Stalingrad to help restore the city after the devastating actions of war."

After 2½ years of that, in her early twenties, Elena met Alexander Kistchinsky—her future husband and the father of

Elena, *left*, at age 9 and her sister, Antonina, age 12, toward the end of World War II (circa 1945) in their home village of Dolinovka.

their two children, Andrei and Yulia. Alexander was a graduate of Moscow State University working in a water fowl reserve, with dreams of going on a biological expedition. They went together to Kamchatka where they had their son in 1959. It was there that Alexander convinced her to learn English, prefacing it by saying, "In order to help me with my research." She began studying English on her own, and then when they returned to Moscow in 1962, she entered the Moscow Institute of Foreign Languages.

"Happily," she says, "it is the best institute of [its] kind, and in five years I graduated knowing English well. It was an evening department; to go to the day department, I did not have enough marks at the exams." Still, she learned English so well that in 1966 she was hired as a translator and interpreter for the prestigious Institute of Geography. "Those years were good and productive," she recalls. "I even entered the postgraduate course and planned to become a true geographer." In 1974, however, she gave birth to daughter Yulia and dropped the postgraduate course to begin working in the Environmental Commission of the IGU.

In 1980, her husband died of cancer, and she was left alone with two students (her 21-year-old son Andrei and his young wife), their newly born baby girl, and five-year-old Yulia. "These 10 years were really difficult," she admits. "One salary was not enough for supporting them all, and I had to work at different international events performing simultaneous translation, which was extremely difficult

Elena was assistant in international cooperation for Professor Vladimir Kotlayakov, *left*, when, in 1989, she first saw the software and convinced him "how effectively this system might be used to advance geographical research and bilateral American-Soviet cooperation in the field of environmental studies, global change research, and forecasting future change."

though it paid rather well." That was in the daytime; she also worked nights translating written texts. "Then came 1991–1992, and my life as a businesswoman began."

When Esri agreed to donate PC ARC/INFO to the Institute of Geography in 1989, getting the software through customs at the airport was a nightmare. Finally, the institute's director at the time, Vladimir Kotlayakov, managed to bring the two heavy boxes through himself—by virtue of his diplomatic passport.

One of Elena's early obstacles involved wrestling with the differences in accounting systems between the West and the Soviet Union. Then she had to deal with importing Esri software products to Russia—before the advent of web GIS. "The customs service was very strict—we had to spend several days in the airport to clear the imported PC ARC/INFO and ArcView and accompanying literature," she remembers. "The third difficulty, and not the least, came in trying to obtain visas for our employees to visit conferences, symposia, and other events organized regularly by Esri in the USA and in Europe," she says.

The first 10 years of our work here were mainly advertising and introducing GIS in Russia and other countries of CIS.

Besides organizing annual conferences, Elena and DATA+ also set up a training center for users, where more than 7,000 specialists have been trained. "We also helped the most advanced customers set up subdistributor offices in different regions of Russia, and we translated into Russian, then published, both the software products and the literature."

With the new century came time to build a new office. The renting of space in different places had become insufficient and too expensive, so they found a plot of land in Moscow and began constructing their own three-story office building.

By 2010, the next step was to create a new company. "We did not want to—and could not—rename DATA+ because by that time it was already established as a respected, well-known company under its own name," Elena says. "We already had many customers who were used to DATA+ and accustomed to working with it. So, we decided to create a new company, including the territory of other countries of the Commonwealth of Independent States. We called the new company Esri CIS, and our market territory included Russia and eight other countries that became independent after the breakup of the Soviet Union." DATA+ became part of the subdistributor network within the Esri CIS market, consisting of 25 companies.

With the cover of his book *Thinking about GIS* translated into Russian in the background, Roger Tomlinson, *center*, and Jack Dangermond, *left*, listen to Elena at the 2006 conference in Russia.

Elena's daughter, Yulia, who started working for DATA+ when she was barely 20, was appointed director general of Esri CIS in 2010 and is still working in that capacity. With Elena's retirement in 2013, her friends and partners Alexei Ushakov and Andrei Orlov respectively became director general and executive director of DATA+. Until her retirement, Elena continued in various capacities at the company she founded, shepherding its focus toward professional services, including application development, web services, implementation, and the production of digital maps. As a team, mother Elena and daughter Yulia coordinated many joint activities between Esri CIS and DATA+, including organizing user conferences, training, language localization of software products, and publication of the quarterly *ArcReview* magazine.

Mother Elena and daughter Yulia.

And today, Elena still appreciates the power of working toward cooperation and a mutual purpose.

These days, [we see] the decision we made 10 years ago, to create Esri CIS, was very wise. Many organizations of various size, having begun working with GIS long ago, cannot imagine stopping, and they continue cooperating with us. ✱

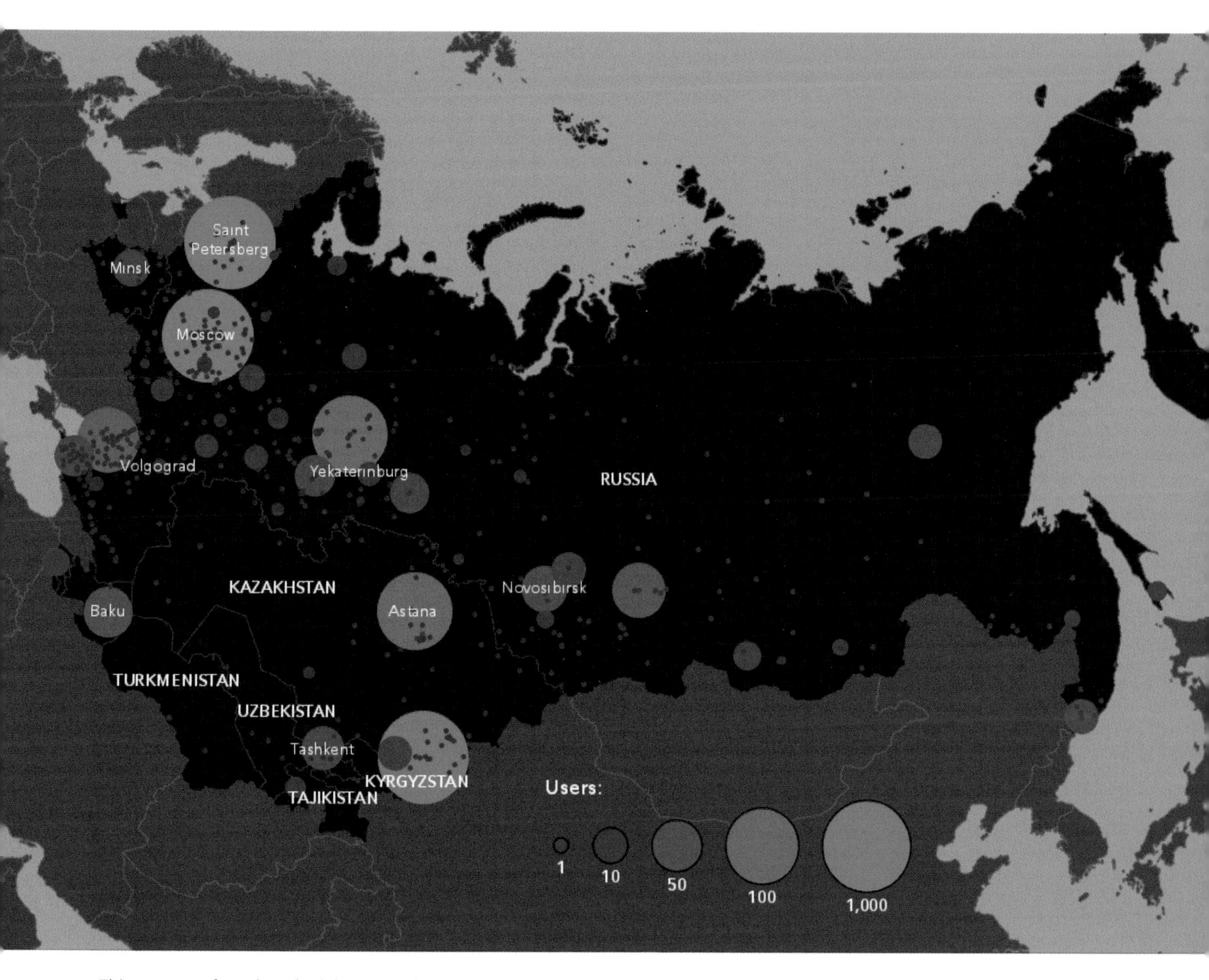

Thirty years after Elena laid the groundwork for using software in post-USSR Russia and nearby countries, international distributor Esri CIS, along with subdistributor and service company DATA+, currently serve many thousands of users (corporations, NGOs, and individuals) in cities all over the region. Map by Olga Serebryannya and Jennifer Bell. Data courtesy of Esri CIS marketing department.

Mary Spence

A cartographer worthy of queen's honors

Mary on the croft with her grandfather.

AS A LITTLE GIRL, MARY SPENCE WALKED DOWN THE DIRT TRACK FROM her grandparents' croft in Kincardineshire to make her way to a Scottish village classroom on the cliffs overlooking the North Sea. This is how her life's adventure started, traversing next to a harbor town, and then winding her way to the city. Her imagination and her curiosity were first inspired by the idyllic Scottish landscape she grew up in and engaged with, and she continued to be guided by her imagination and curiosity as she studied the natural world, drawing maps first with colored pencils and later with computer software. Mary dedicated herself to excellence as she continued her journey and made her way to England, where she still works as an award-winning cartographic designer, author, and teacher. Along the way, she made a stop at Buckingham Palace where she was honored with a Member of the Most Excellent Order of the British Empire (MBE) award for Services to Cartographic Design.

As an only child in the Highlands, Mary spent most of her time outdoors with nature and animals. Her primary school teacher would occasionally walk the students to the beach and let them explore the rocks in the bay. Mary was always asking questions and

wondering how the world worked, such as whether the same cloud pattern she saw in the spring would appear again the next spring. In secondary school, geography was her favorite subject. Her teacher, Mr. Gray, would start a class by drawing a map of the country being taught, and then annotate it with the various features he discussed. He introduced group projects where students went off-curriculum to investigate a geographical concept and submit their research at the end of the term. Mary remembers drawing a world map depicting the route of Sir Francis Chichester in real time as he sailed alone around the world in a 55-foot yacht in 1966–67.

Mr. Gray would take his students on field trips, piling them into his dilapidated car to go exploring. They studied the rock conglomerates on the beach at Stonehaven, the line of the Highland Boundary Fault just north of town, the Roman camp at Raedykes, the cliff formation at Dunnottar Castle (see photo), and the story of the smuggling of the Honours of Scotland (the Scottish Crown Jewels) from the castle to be hidden in a church to avoid them falling into the hands of the English. This hands-on learning inspired Mary.

The landscape that inspired Mary's imagination.

Mary went on to earn a degree in geography at the University of Aberdeen and thought she'd become a geography teacher like Mr. Gray. Her cartography tutor, Mike Wood, however, suggested that Mary continue her education in cartography at the University of Glasgow. He had opened Mary's eyes to a wide variety of maps, and he saw her enthusiasm and recognized her potential. When Mary realized that she could possibly have a career making maps, she was eager to learn more. At Glasgow, she took a course by John Keates, a renowned teacher and innovator in the design of thematic and recreational mapping. Through this instruction, Mary would learn amazing possibilities with maps.

Mary started her career as a cartographic editor with Pergamon Press in Oxford, England. Mary saw that there was a pervasive thought in the industry that it wasn't worth training a woman because she would leave to start a family. To counteract that widespread belief, Mary worked for 10 years to establish herself before she had a child, Benjamin, in 1982, and then returned to work immediately thereafter and continued to work full time throughout her career. She had a daughter, Samantha, in 1987. Men were typically paid more than women because they were the breadwinners. But in her family, Mary was the breadwinner, and her husband stayed home to take care of their children. Since this wasn't a "normal situation," Mary had to speak up and continually fight for increased pay.

Mary then began working for David Fryer & Co, which became GEOprojects (UK), where she was appointed chief cartographic editor and then general manager. During her tenure, the company won many awards for cartographic excellence. While working as chief cartographic editor and preparing an atlas of the United Arab Emirates (UAE), Mary traveled to the University of Al Ain in the UAE, her first time in the Middle East. The professor of geography she was meant to work with was reluctant to collaborate with "a girl from England" but soon realized that she was a capable project manager and a valuable addition to the team.

Determined to get things right, Mary is one to never accept second best. If a map can be improved, she feels she must work to improve it. When she reaches a point where she knows she is struggling to understand a problem, she steps aside and starts asking questions, seeking advice until she comes up with a solution. She keeps learning from "outstanding" mapmakers. In Oxford, she visited the Bodleian Library to study foreign atlases, especially those by the Dutch and German mapmakers whose color combinations looked so different from what she had studied in school and university. She learned innovative map design from David Fryer, who seemed far ahead of other publishers at the time, and Bob Hawkins, whose Royal Engineers training ensured that Mary had a basic grounding in the serious discipline of mapmaking. His exacting standards laid the foundation for Mary's future success in the industry.

When Mary started her career, maps were produced in two distinct phases by people in two distinct roles. Cartographic editors (usually with an advanced degree) researched, compiled, and designed the map content, and then sent it to a draftsman (without a degree) who handled the technical part of the map drawing. When computers came to the workplace, it was assumed that they would be the tools of the draftsman, and many editors left the industry altogether.

Mary saw the introduction of computers as an exciting development and was determined not to be left out. When Mary joined Global Mapping, another map publisher, as project manager, she finally tackled the program Adobe® Illustrator® after a few frustrating years trying to get a draftsman to carry out her changes. With the help of her new colleagues, she learned how to do basic design work on the computer. She continued to practice and explore software tools, but once she was introduced to GIS, she chose to go "back to school" and get extra guidance from Tim Rideout, another mentor and director of XYZ Maps, a company specializing in digital cartography, aerial imagery of Scotland, and GIS software. Becoming a proficient success, she won an Ordnance Survey OpenData Award in 2014.

I came a long way from [being] a young child whose first writing material was a slate and a piece of chalk.

Although Mary was always part of a team that won awards for cartographic excellence, it wasn't until she was able to create her own maps from scratch that she was given free rein to design and develop concepts of world geographies such as the *Dynamic World* and the *Environmental World* maps, both of which won major awards.

Mary working on the *Environmental World* wall map, winner of the Stanfords Award for Printed Mapping and the coveted British Cartographic Society Award in 2011 and the Best Wall Map 2012 International Map Trade Association award.

Mary uses a wide variety of software for different map scales and design requirements. Sometimes it can be hard to remember how she achieved certain effects on previous maps, which can be frustrating. She puts it in perspective, though.

> **It is just another tool—a sophisticated and time-saving tool, which frees the mind to explore aspects of creativity that the slow production methods of the past may have prohibited. The whole point of map design is deciding what would work, then making it happen, not taking what is offered by default. To my mind, default is merely a suggestion to get you started [and] not necessarily the best option in your particular situation.**

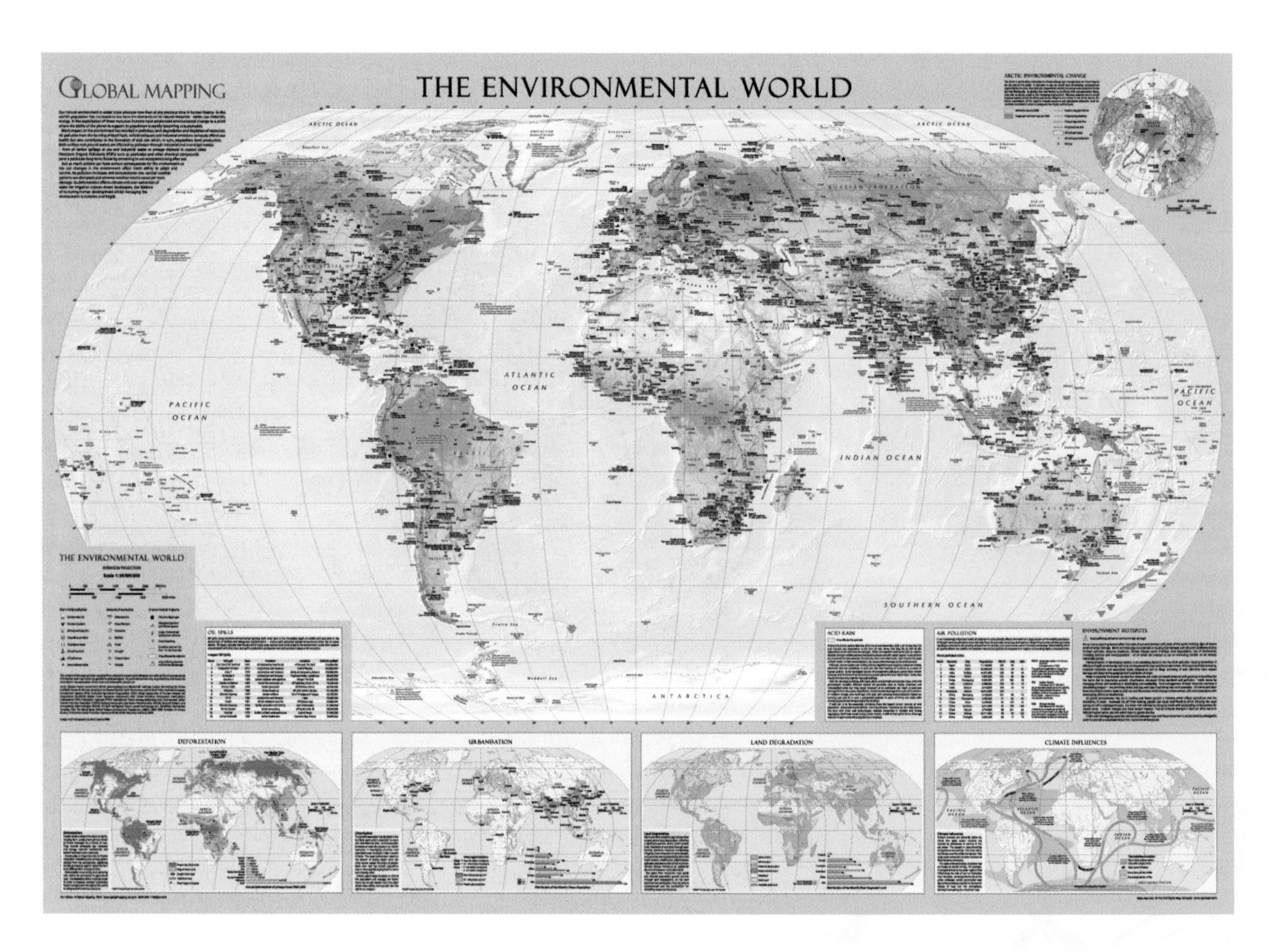

Mary's multi-award-winning educational wall map *Environmental World* highlights the threats human activity places on our planet. The judges who awarded this map the British Cartography Society Cup for Best Overall Map commented that it worked well on all levels, demonstrating good mapping, clarity, symbology, and use of color—plus, it was educational.

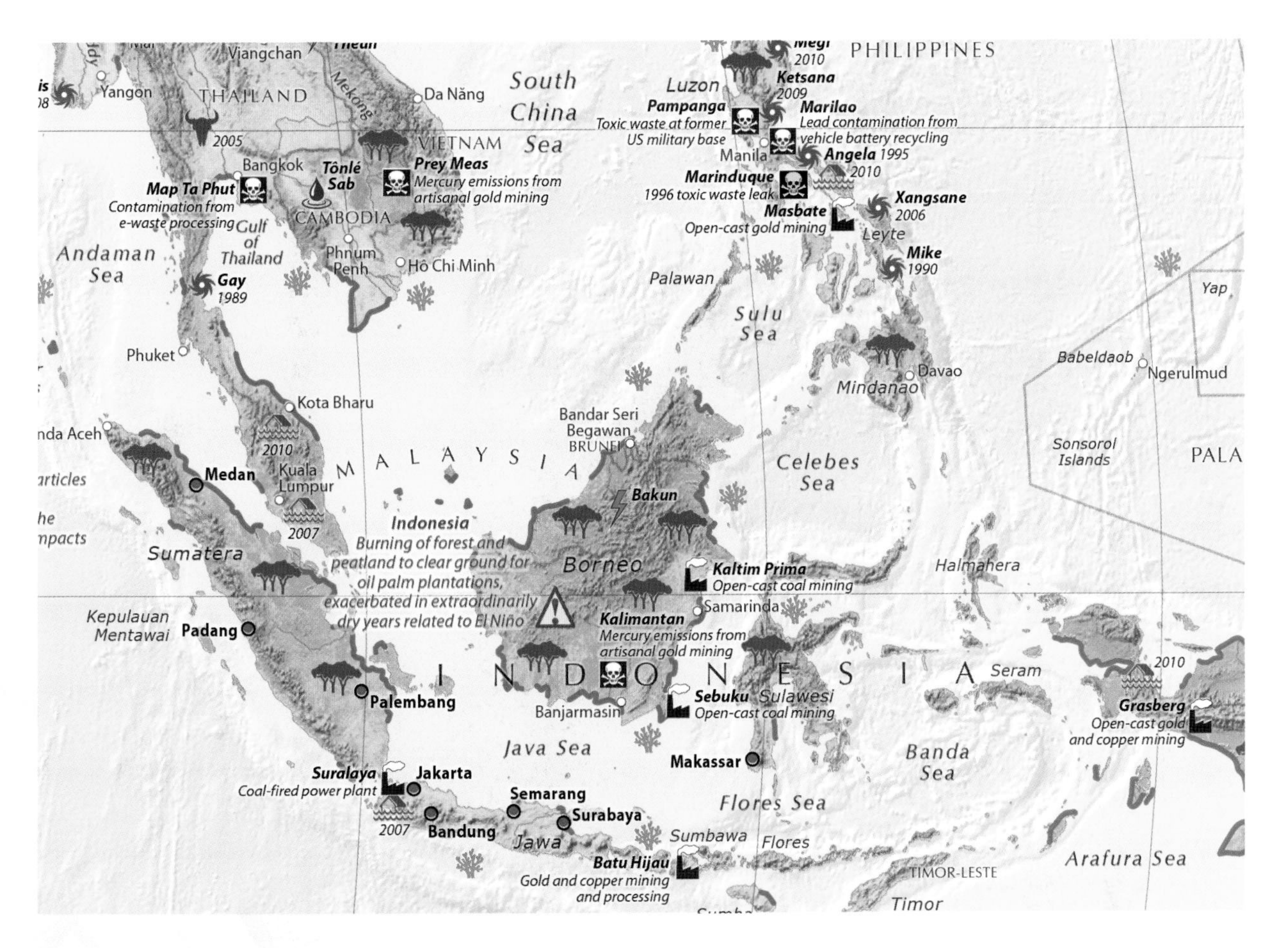

An extract showing details of Indonesia and environs. Beating out *National Geographic* and Ordnance Survey among others, Mary was delighted to win the British Cartography Society Cup and commented, "It is a great honor to win these awards in the face of such giants in the world of maps. I can now allow myself the whole weekend to celebrate before it's back to the drawing board first thing Monday morning."

Mary holds her MBE medal that can only be worn at official events when the monarch or her representative is attending. Mary 'hired' this perfect, posh hat especially for the day. Her family attended the investiture with her. And the best part of Mary getting her MBE—how proud her mum is.

The highlight of her career came in November 2003 when she received a letter from 10 Downing Street, asking "if it was agreeable to her that the prime minister submit my name to the queen with a recommendation that Her Majesty may be graciously pleased to approve that I be appointed a Member of the [Most Excellent] Order of the British Empire." Shocked, Mary read and reread the letter until it became real.

Needless to say, it was agreeable to me. Then it had to be kept top secret until the New Year's Honours List becomes public, but my family were there when I opened the letter and my son asked if I had seen a ghost! And, of course, I told my mum.

The day itself was a blur of excitement from getting into the taxi at the hotel and saying to the driver, "Buckingham Palace, please," then the investiture itself which I can scarcely recall, followed by afternoon tea at the Ritz. Oh, and the hat! It was Prince Charles who presented my MBE, and he asked if maps were not all made by computers nowadays, to which I replied that the computers needed someone to tell them what to do, or something equally corny.

Mary recalls that "for weeks afterwards, I could not recall the detail of the day, but I visited the Palace as a tourist the following year and retraced my footsteps."

Mary received many congratulations from the cartographic world, including a personal letter from the director general of the Ordnance Survey thanking her for her enthusiasm and promotion of the discipline.

However, as a career move, it was a nonstarter. One chap went so far as to say I was now unemployable–"no one is going to take on someone with an MBE because they will think you are after their job," he said. Maybe so, but I was now at a stage in my career where the recognition for my skills and ability has kept me engaged in producing a wide variety of maps with there being no sign of demand for my services slowing down.

Mary was president of the British Cartographic Society from 2006 to 2008 and sits on the council as a design consultant. As an active member of the Programme Committee, she is involved with the annual symposium, Better Mapping seminars, and Restless Earth Workshops for secondary schools. She was also a guest lecturer at Oxford Brookes University when it ran its cartography courses and in 2017 spent an afternoon with topographic science students at Glasgow University talking about map design and assisting them with their dissertation maps.

I firmly believe that interest in maps can start very early. I visit primary schools telling children about the wide variety of different maps that there are and getting them drawing their own maps.

Mary is a Fellow of the Royal Geographical Society and has coauthored and published two editions of *Cartography: An Introduction*—a practical guide to designing better maps, aimed at those starting out on the mapmaking journey.

Mary, *center*, works with primary schoolchildren to draw maps.

Mary hopes to influence next-generation computer technicians who are making maps. "GIS has all the tools required to make superb maps, and cartographic principles can enhance GIS output to provide a more meaningful map that communicates its message efficiently and effectively. ... Cartography will become a niche expertise. ... The results can be catastrophic, but so long as there are people who find those as acceptable, then they will prevail," she says. "However, I think there will always be a place for a well-designed, informative, and attractive map, whether it be in print, online, or on a smartphone."

She encourages young women to pursue their dreams to the best of their abilities.

Fight for what you know is right, and do not undersell yourself. If you don't already possess a skill, learn it. Keep trying and keep improving. You will get there. ✷

Kathryn D. Sullivan

Meeting the challenge, from NASA to NOAA—and beyond

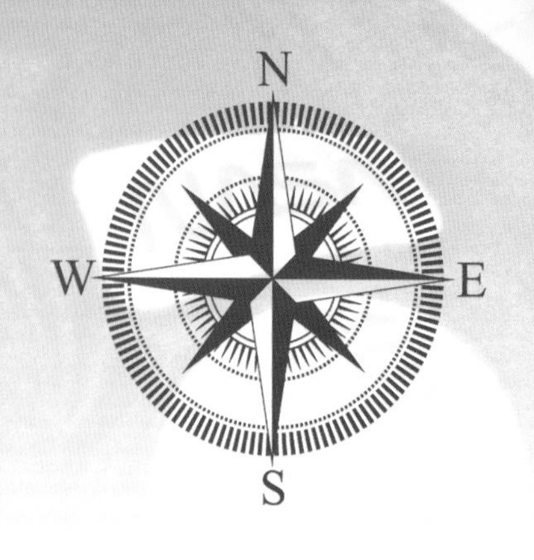

DURING HER PREMIER SPACE FLIGHT IN 1984, DR. KATHRYN D. SULLIVAN, America's first woman to walk in space, spent three and a half hours outside the space shuttle *Challenger* 140 miles above Earth, demonstrating the feasibility of refueling a satellite in flight. With that feat, she became the first US woman to perform an extravehicular activity (EVA), or spacewalk.

She was 33 years old at the time of this spacewalk. Out of nearly 9,000 applicants, NASA had chosen her, at age 26 in 1978, to be one of the first six women American astronauts. She was a member of NASA's storied astronaut class of 1978, which also included the first African Americans and the first Asian American. Instead of selecting only pilots and those with a military background, NASA now planned to employ many scientists in the role of explorers.

Kathryn served on three space shuttle missions, aboard *Challenger* in 1984, *Discovery* in 1990, and *Atlantis* in 1992, and was inducted to the Astronaut Hall of Fame in 2004.

Kathryn turned six the day before *Sputnik* was launched. This was the Cold War era of the space race between the US and the Soviet Union, when all eyes turned to the skies. Growing up in Woodland Hills, California, Kathryn pored over paper maps and the pages of *Life*, *Look*, and *National Geographic*. She thirsted to join the adventurers and explorers pictured in exotic places. Her dream of

Kathryn on her first space flight, catching sight of Earth from space.

NASA (Oct. 5, 1984).

exploring the world was driven by a broad curiosity and an interest in mapping, rather than an interest in any one scientific discipline.

She earned a bachelor of science in earth sciences from the University of California, Santa Cruz, and, five years later, a doctorate in geology from Dalhousie University in Halifax, Nova Scotia. Two of her freshman-year college professors were key influences, opening her eyes to how adventurous and fun the life of a scientist can be. They supported her as she changed her major from foreign

The patch from the space shuttle program, 1981–2011, NASA's fourth human space flight program. Its official name, Space Transportation System (STS), describes its function, to carry crew and cargo from Earth into orbit and back again, in a reusable vehicle. NASA.

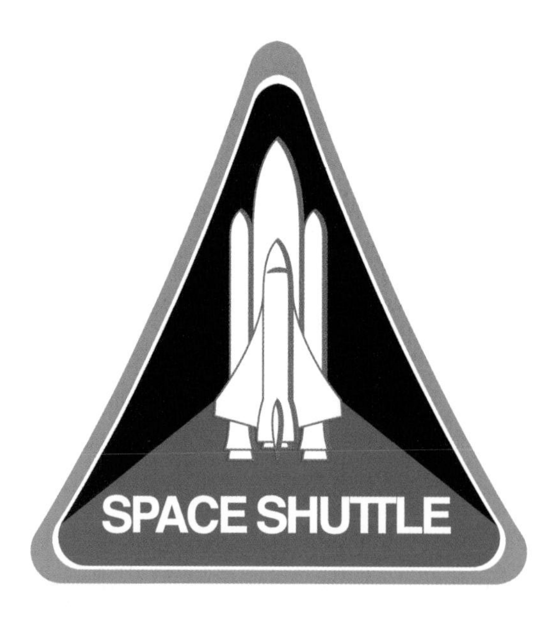

languages to earth sciences. She was finishing up her PhD when she learned that NASA was recruiting scientists and engineers to become astronauts in its new space shuttle program. Kathryn recalled:

To see the earth with my own eyes from orbit—that wasn't an opportunity I was going to pass up. You could have told me the odds [of being selected] were one in 200 million and I still would have applied, because that means they're not zero.

Kathryn flew two more missions with NASA after her 1984 spacewalk. Her second flight was aboard the space shuttle *Discovery* in April 1990, on the mission that deployed the Hubble Space Telescope. This flight set an altitude record of 380 miles and orbited Earth 76 times in 121 hours. Hubble became one of the most successful astronomical instruments ever created. For her third and last flight, aboard *Atlantis* in 1992, Kathryn served as payload commander on the first spacelab mission dedicated to NASA's Mission to Planet Earth. The scientific instruments on this mission made chemical and physical measurements of the atmosphere that helped scientists understand the processes of climate change.

After logging more than 532 hours in space, Kathryn's desire began to shift from collecting data in space to working on the ground. She left NASA in 1993 to accept the post of chief scientist for NOAA in Washington, DC, becoming the second woman in that role. She followed immediately behind her longtime friend Sylvia A. Earle (also featured in this book).

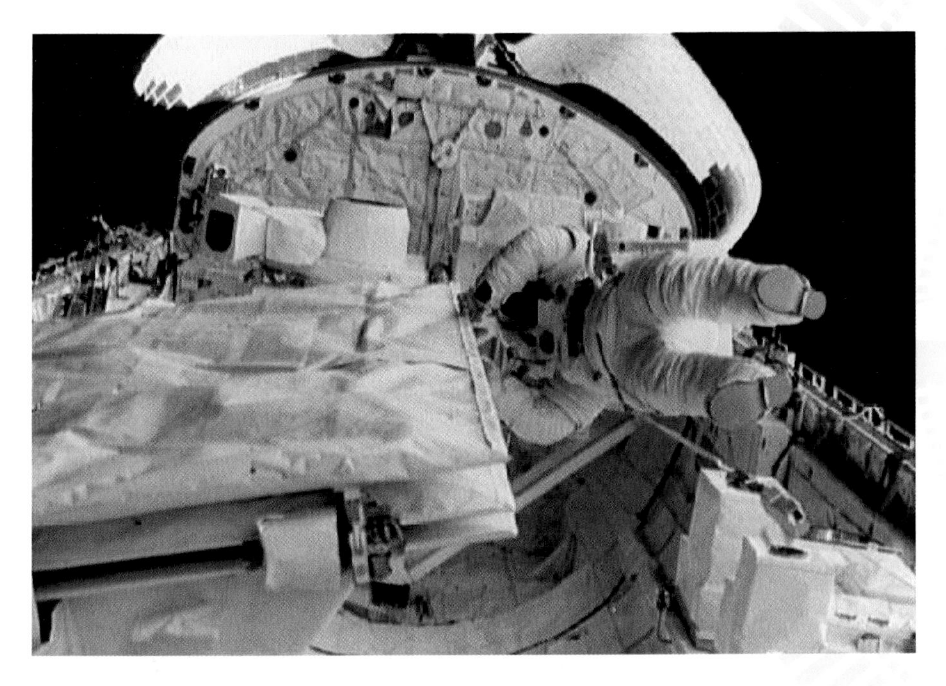

As the first American woman astronaut to work outside a spacecraft, mission specialist Kathryn checks the latches on the SIR-B radar antenna in the *Challenger*'s open cargo bay during her EVA on October 11, 1984. Just beyond her helmet is the orbital refueling system, which she and crew mate David Leestma operated to show that a satellite could be refueled in orbit. Film footage from this STS-41-G mission appears in the IMAX movie *The Dream is Alive*. NASA.

Upon Kathryn's being chosen as one of *Time* magazine's 100 most influential people of 2014, John Glenn pointed out that she served aboard the mission that deployed the Hubble Space Telescope: "That role in helping humanity look outward has not prevented her from looking homeward. The planet is suffering increasingly severe upheavals, at least partly a result of climate change–droughts, floods, typhoons, tornadoes. I believe my good friend Kathy is the right person for the right job at the right time." NOAA.

Enamored of space, Kathryn wanted to bring the power of that perspective down to earth to help inform the many decisions people make every day about the environment. This desire was a major influence in her decision to accept the nomination as NOAA chief scientist.

"To me, the experience of being in space led me to understand that the most pressing challenges of my world are, in fact, those that face us all collectively," she says.

Circling the world every 90 minutes for days on end, it became clear to me that all of us on Earth are inextricably linked to each other as humankind. There is no "I" or "them"–we are one, and our very existence is completely dependent on the workings of that "little blue marble." And so, I came to believe that I owed it to myself and the people who put me in space to make good on that unique point of view. It was this newfound perspective that led me to leap into my next professional adventure, putting scientific understanding of the earth to work for all of us. So when I returned to Earth, I built–not found, but built!–a career focused on achieving that outcome.

On the sixth flight of space shuttle *Challenger* (October 5–13, 1984), the first shuttle mission with two women, are Kathryn (*left*, on her first flight) and Sally Ride (the first American woman in space, on her second flight).
NASA.

About five years later, Kathryn was debating with herself whether to remain in government or make another, even more significant career shift.

I really wanted a chance to shape, design, and lead something. When I thought more carefully about the leadership challenges I'm really interested in, I realized that you don't have the same kind of latitude in most government programs that you do if you're actually creating [something].

What she wanted was "to conceive, design, formulate, organize, deliver, muster the resources" for some worthy endeavor. But what?

She made a list of people she knew who were successful outside government, and, one by one, she took them out to lunch to learn more about their work. "I bought about a dozen people lunches: 'How does your world work? What are the greatest things? What are the best parts of it? What are the ugly parts of it? What are the huge challenges? What drew you into it? What keeps you in it? If, at the end of this term, two or three years from now, I was to parachute into this world, what do you think the emerging, really interesting challenges and opportunities would be?' "

Kathryn was in her early 40s at this point in her career, and midway into her chief scientist role, but she was always one to look a step ahead.

If you want to add a really different next chapter in your professional life, the early 40s is a good time to do that. You've got good running room ahead of you. [For me] it seemed like the right, still fertile, vibrant point in time to launch. If you're going to launch a whole new chapter, jump in now at this still fresh, vibrant phase, and go for it and see what it becomes.

Kathryn chose to work for a nonprofit that gave her the latitude to become the creative, hands-on, innovative leader she wanted to be. She served as president and CEO of COSI (Center of Science and Industry) in Columbus, Ohio, from 1996 to 2006, at a time when the science center sought a new beginning as an innovator of inquiry-based science learning resources for the classroom. Under her tenure, a new $125 million headquarters was built, and COSI became the number one science center in the country.

Kathryn's next career step was serving as the inaugural director of the Battelle Center for Science and Technology Policy in the John Glenn College of Public Affairs at The Ohio State University. She returned to government work in 2011, when President Barack Obama appointed her assistant secretary of commerce for Environmental Observations and Predictions in NOAA. NOAA's mission aligns perfectly with the impulses that moved Kathryn to leave NASA in 1993: to understand and predict changes in the earth's environment and share that information with others so they can make wise decisions. Two years later, she was appointed to lead NOAA as the undersecretary of commerce for oceans and atmosphere and as NOAA administrator.

At the Esri Ocean GIS Forum at the Esri Conference Center in Redlands, California, in 2017, Kathryn spoke of the tremendous challenges and opportunities that lie in store for GIS and the study of the planet.

"The space age has helped us understand how the planet really works," she says.

Today we look at the earth in countless ways. Countless sensors–on land, in the oceans, mounted on satellites–are taking billions of data points about this planet every day, and we don't even think about them anymore. Step back from this moment and realize that this is new in our lifetime. All this data requires more resources to make sense of it and understand it, and to put it to good use for the benefit of the planet.

Kathryn's hope for the future:

Our world needs more, not fewer, bright and energetic people to become scientists, both to advance the frontiers of knowledge and to connect science to society. I would hope to see more women taking this path. It may not be the easiest career path you can choose, but it will give you an intellectual foundation and reasoning capacity that will serve you well in every aspect of personal and professional life. It will also open the door to more, and more varied, career options that pay significantly better than other work. ✷

Nancy Tosta

Setting standards and seeking consensus

Nancy continues to work to help others understand the systems we inhabit and the potentially irreversible changes we are making in them.

NANCY TOSTA HAS SPENT A CAREER ADDRESSING MANY OF OUR MOST vexing environmental and social challenges, working across sectors, disciplines, and levels of government to find solutions. Educated as a scientist, she became an early adopter of GIS in California state government in the late 1970s. At that time, the various computer mapping software systems were incompatible, and digital spatial data didn't exist.

In the early 1990s, Nancy helped launch and institutionalize federal geospatial data sharing, the National Spatial Data Infrastructure (NSDI), and metadata standards as staff director of the Federal Geographic Data Committee (FGDC) and a special assistant to Secretary of the Interior Bruce Babbitt. More recently, she is serving her local community (Burien, Washington) as a city councilmember, addressing issues ranging from affordable housing to climate change. She continues to seek ways to use her analytical and consensus-building skills to serve the public.

Growing up in Connecticut in the 1950s, Nancy played in the woods and the swamp behind her house, collecting worms and bugs until the land turned into a housing development. She discovered a deep interest in nature and read constantly, including many adventure series such as The Hardy Boys and science fiction. Her family supported her as she started to write stories about space travel and extraterrestrial life. Her sixth-grade teacher allowed her to set up science experiments growing brine shrimp in the school basement and observing the shrimp under a microscope. Nancy says, "I was hooked on understanding science. I don't know why he made these exceptions for me, but I believe he fits the profile of a 'great teacher.' "Her seventh-grade "new math" teacher similarly paid attention to her keen interest and abilities, and when she looks back, she realizes that the early encouragement of male teachers made a difference in boosting her confidence, and thus her success.

It was mostly that they treated me as if I was capable, gave me confidence, and perhaps made me feel special. It was being recognized, acknowledged, and complimented by the male teachers–especially in science and math, where it was mostly boys–that had the most impact.

At the University of California, Berkeley, in the early 1970s, Nancy studied soil science and plant nutrition, earning bachelor of science and master of science degrees. She loved the "full systems" thinking that knowledge of soils required, including understanding geology, botany, climatology, geomorphology, and biology. While at Berkeley, Nancy worked for the Space Sciences Laboratory analyzing the signatures of soils and vegetation for NASA research on spectral bands to be included in the initial Landsat satellites. After graduation, Nancy worked for the California Department of Forestry and Fire Protection (CDF) as a soil scientist and in 1978 was chosen to lead a project working with NASA to map the vegetation of California using Landsat data.

In the late 1970s, several California state agencies were exploring computer mapping using different software. Few people understood coordinate systems, map projections, or scale, let alone computers—and there was no internet and very little digital spatial data. Nancy recalls spending hours running a mouse over lines on paper maps on a digitizing table to create digital maps of hydrology, vegetation, and statewide landownership patterns. In 1980, Nancy chaired the California Computer Mapping Coordinating Committee, which was formed because she and others knew that sharing data could be helpful.

It was a group of us from six to 10 different agencies who realized we were all working with a new technology and all needed the same data–roads, boundaries, watersheds. We wanted to figure out how to minimize redundancy and share data, even though our systems had a very hard time "talking" with each other.

The committee argued over which software was best and bemoaned the lack of metadata, which prevented them from using data from each other's department or agency. In the early 1980s, the CDF purchased its first minicomputer (Prime) and its first GIS software (ArcInfo®), and Nancy was responsible for all software and hardware installation, plus data capture. All of this experience led Nancy to see the powerful potential of GIS.

In 1988, Nancy was hired to develop and run the GIS Laboratory at the Teale Data Center, at that time California's largest data processing center, maintaining the data for many statewide offices such as the Department of Motor Vehicles and state personnel files. Nancy helped launch a statewide digital data library to enable more standardized data sharing. She recalls endless debates on whether spatial data should be freely available or sold to recover costs. Nancy was an advocate of the free use of data. As a deputy director of Teale, Nancy believed it was important to ask users about their needs rather than dictate to them which software they must use. Nancy began writing a monthly column in a trade magazine addressing GIS issues, data sharing, education, and more, developing a fan following.

In 1992, Nancy was hired to promote data sharing among federal agencies, working as staff director of the FGDC. Leading a staff housed in the US Geological Survey headquarters outside Washington, DC, Nancy helped evolve the concept of the NSDI to include data standards, a framework of consistent spatial data, and partnerships between agencies. In 1994, Nancy was the primary architect and shepherd for passage of Executive Order 12906, setting legal policy on "the technology, policies, standards, and human resources necessary to acquire, process, store, distribute, and improve utilization of geospatial data." The signing of this executive order raised global awareness of the role governments might serve in promoting data development and sharing and launched efforts in many other nations, ultimately supporting development of a global spatial data infrastructure.

The signing of Executive Order 12906 made many other countries take notice and launch similar endeavors. It was a watershed moment that led to enhanced consciousness of the potential for, and value of, sharing geospatial data digitally.

[Left] In 1995, Nancy (no. 8) was the only woman among 10 individuals identified as one of the Heavy Hitters of GIS in *GIS World* magazine.

[Right] Nancy is seen serving her community as a councilmember, trying to find common ground in increasingly divided times.

In 1996, with an interest in exploring policy work in other areas, Nancy moved to the Seattle area to work for the Puget Sound Regional Council as director of growth strategies and modeling, overseeing work in growth management, GIS, and transportation modeling. She served as president and on the Board of Directors of the Urban and Regional Information Systems Association (URISA) from 1999 to 2000. In 2002, URISA gave her its Leadership Award, recognizing her contributions to GIS and URISA.

Since 1999, Nancy has worked as a consultant to public agencies and foundations on spatial data strategies, climate mitigation and adaptation, environmental health, salmon recovery, rural economic development, and food systems change. Although she retired in 2016, she continues to serve the public, now in her second term on the Burien City Council. Burien is a town of 52,000 diverse residents (ethnically and socioeconomically), experiencing growing pains from the explosive growth of Seattle, 10 miles north. As a local elected official, Nancy focuses on many issues, including city budgeting and funding; homelessness, housing affordability, and tenant rights; quality of life and environmental protections, exacerbated by airport proximity; land use, zoning, and comprehensive planning; public safety, gangs, and police contracts; community engagement and diversity; and infrastructure development and maintenance. In 2017, she received a Public Servant of the Year honor from Discover Burien. Since 2017, she has served on the Board of Directors of Tilth Alliance, a statewide organization in Washington, working to promote healthy eating for all and profitable production for sustainable farmers in the state. She is serving as board chair in 2019–2020.

Nancy is a strong advocate for education, especially at an early age. She would love to inspire young people to study science as her teachers did for her. Working with a local environmental center and a middle school, Nancy encourages kids to go outside, grow food, experience the beach, and touch nature in any way they can.

I fear today we stifle the natural curiosity of many children with too much structure and scheduling. Screens are dominating the lives of our children, and that scares me.

Nancy encourages all young people but would especially like to see more young women take an interest in science to understand how the world works: "Without understanding how the world works, you will have little chance of changing it." Her advice to women professionals:

Don't be afraid of starting over. I've done that–jumping into situations where I didn't know the job nor was I a known quantity. It takes a bit of ego swallowing, but it's really rewarding to continue to learn and master new fields. ✷

Madison Vorva

Taking her seat at the table

BEFORE GRADUATING FROM POMONA COLLEGE IN 2017 WITH A BA IN environmental analysis at the age of 21, Madison Grace Crimmins "Madi" Vorva wrote a thesis that she described as "a culmination of the last decade of my life's work." Indeed, at a young age, Madi had already accomplished a lot on the world stage.

Madison Grace Crimmins "Madi" Vorva.

Now at age 23, Madi is an avowed environmental conservationist who has succeeded in catalyzing major global change—as a child and as a teenager, she campaigned to get the Girl Scouts to stop using unsustainable palm oil in their cookies. Through research and perseverance, she convinced the Kellogg Company, leading producer of cereal, cookies, and crackers with its Kellogg's brand, to agree to a worldwide deforestation-free policy for all its products.

She was among the first United Nations Forest Heroes in 2012, at age 17, and was celebrated at the *Teen Choice Awards*. She joined the Board of Directors of the Jane Goodall Institute (JGI) in 2015. She hosted the television program *FabLab* to show kids—especially girls—how fun and important science and technology are, and she developed a GIS-based curriculum for high school students to encourage spatial thinking.

From age six, she's fought practices that send wild animals to the brink of extinction and has advocated for sustainable ways to preserve land and habitat—for indigenous people as well as wildlife.

Madi remembers when she was six, falling in love with orangutans at the Fort Wayne Children's Zoo in Fort Wayne, Indiana. On another field trip, in second grade, she saw *Jane Goodall's Wild Chimpanzees,* and decided, "I want to be her when I grow up!" Says Madi: "I wanted to save animals, like Dr. Jane," so she went home to her own Michigan neighborhood and created a "save the endangered species" club in elementary school.

Madi at age 14 with her hero, Jane Goodall, *left*.

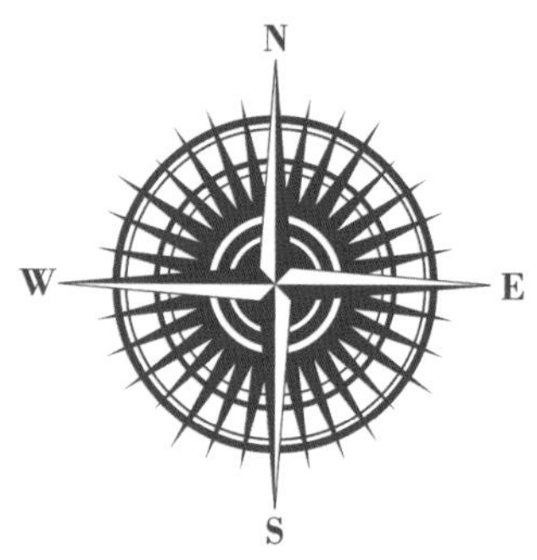

She and friend Rhiannon Tomtishen, both Girl Scouts, immediately set out to learn as much as they could about primates. They discovered that a major threat—especially to the survival of orangutans—is the loss of habitat from deforestation to make room for oil palm plantations. Alarmed at learning that palm oil is used in about 50 percent of packaged American grocery store products, the two girls hosted palm oil–free bake sales and created Orangutan Week at their school.

Then one day, at age 11, Madi flipped over a box of Girl Scout cookies and saw that palm oil was an ingredient. "That was really shocking because I'd sold the cookies since I was in first grade," she says. What she did next is now described in her undergraduate college thesis:

After sending an email to the manager of product sales outlining its moral obligation to stop using palm oil, I assumed the organization [Girl Scouts USA] would change its ways immediately. This wasn't the case, and I spent my adolescent years growing Project ORANGS, a campaign for sustainable palm oil in Girl Scout cookies, from a middle school initiative to an international platform.

Girl Scouts Rhiannon, *left*, and Madi, *right*.

Madi and Rhiannon decided to sell magazines instead of cookies. Planning to boycott the baked goods until the Girl Scouts stopped using palm oil in them, they started a petition on Change.org to get others to join in. They attended a conference for Dr. Goodall's Roots & Shoots, a youth-led program in nearly 100 countries that inspires kids to create community action projects, to see if Jane would sign the petition. She didn't right away.

She actually changed the language on the petition. [The gist of] our original version was, "I'm not going to buy Girl Scout cookies until the palm oil is taken out," and Jane Goodall rewrote it [to say], "unless it's from a sustainable plantation."

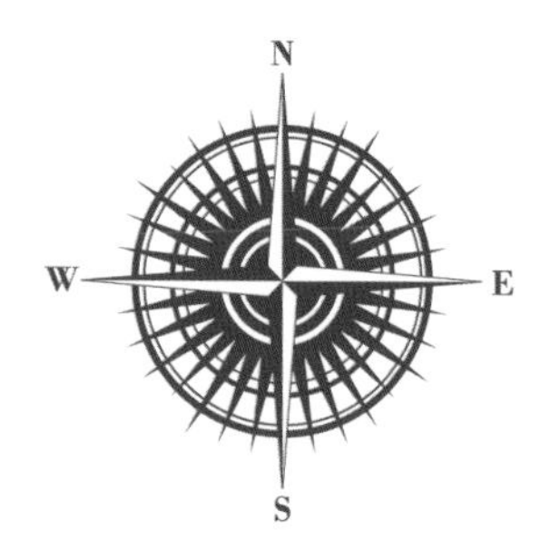

With that wording, more than 70,000 people signed the petition, including Jane, whose input got Madi thinking about *sustainability* and what that meant. The girls reached out to the Rainforest Action Network, the Union of Concerned Scientists, and other organizations to get them to partner with them on the campaign. Through such collaboration, Madi started to realize that oil palm farming not only threatens orangutans but also is linked to the displacement of indigenous communities and to enslaved child labor.

Part of the Girl Scouts' mission is to advocate respect for the earth and encourage using resources wisely, so the two Scouts remained members of the organization throughout their campaign and worked to change it from within. They started a Rainforest Hero badge and sponsored art projects to help girls learn about the palm oil issue and its connection to deforestation. Soon, the girls were running a nationwide campaign.

When Girl Scouts USA agreed to meet with Madi and Rhiannon (then 14) at its headquarters in New York, it was big news in 2009—on the front page of the *Wall Street Journal* and in *Time* magazine as well as on ABC, NPR, and CBS. "It was all over the place my sophomore year of high school," Madi recalls. The Girl Scouts chose to offset its use of palm oil by purchasing certificates of credit from the GreenPalm program.

"That was really exciting, but it was also discouraging because they didn't make sure their palm oil was from sustainable sources," Madi said.

So she decided to go bigger. The Kellogg Company, based in Madi's home state of Michigan, owns one of the two companies that bake Girl Scout cookies. So Madi and her team—with support from hundreds of thousands of people—began communicating with Kellogg's to persuade the company to use deforestation-free palm oil. Not only did Kellogg's comply by announcing in 2014 that it would implement a deforestation-free policy for all its products, but Wilmar International, which trades 45 percent of the world's palm oil, agreed to adopt a landmark deforestation-free policy as well. By 2015, 90 percent of palm oil traded internationally was bound by deforestation-free policies, and most American companies now have a sustainable palm oil policy in place.

Meanwhile, Madi was busy doing other things as well. She studied land rights in Colombia, traveled to Cambodia to make a short documentary on a politician fighting for victims of land grabs, and observed how palm oil farming affects local communities in Malaysia and Singapore.

At 19, she was chosen as the youth speaker for the gala celebrating Jane Goodall's 80th birthday. The next year, Madi attended the 2015 United Nations (UN) Climate Change Conference in Paris and shot videos explaining to kids what the UN does.

In college during an introductory course in GIS technology, Madi geocoded the petition signatures from her original campaign targeting the Girl Scouts' use of palm oil.

It just blew my mind to be able to say for the first time, "This is where everyone was from, this is the most concentrated Zip Code for petition signatures." I wish I had known that was out there when I was younger because you can visualize your impact so much easier. If we had had a map of social media when the topic was trending, that would have been a great tool to show that this [campaign] was producing real results.

During her college years and after graduation, Madi continued her work by taking trips to various parts of the world with conservation groups. "The people who should be able to decide what

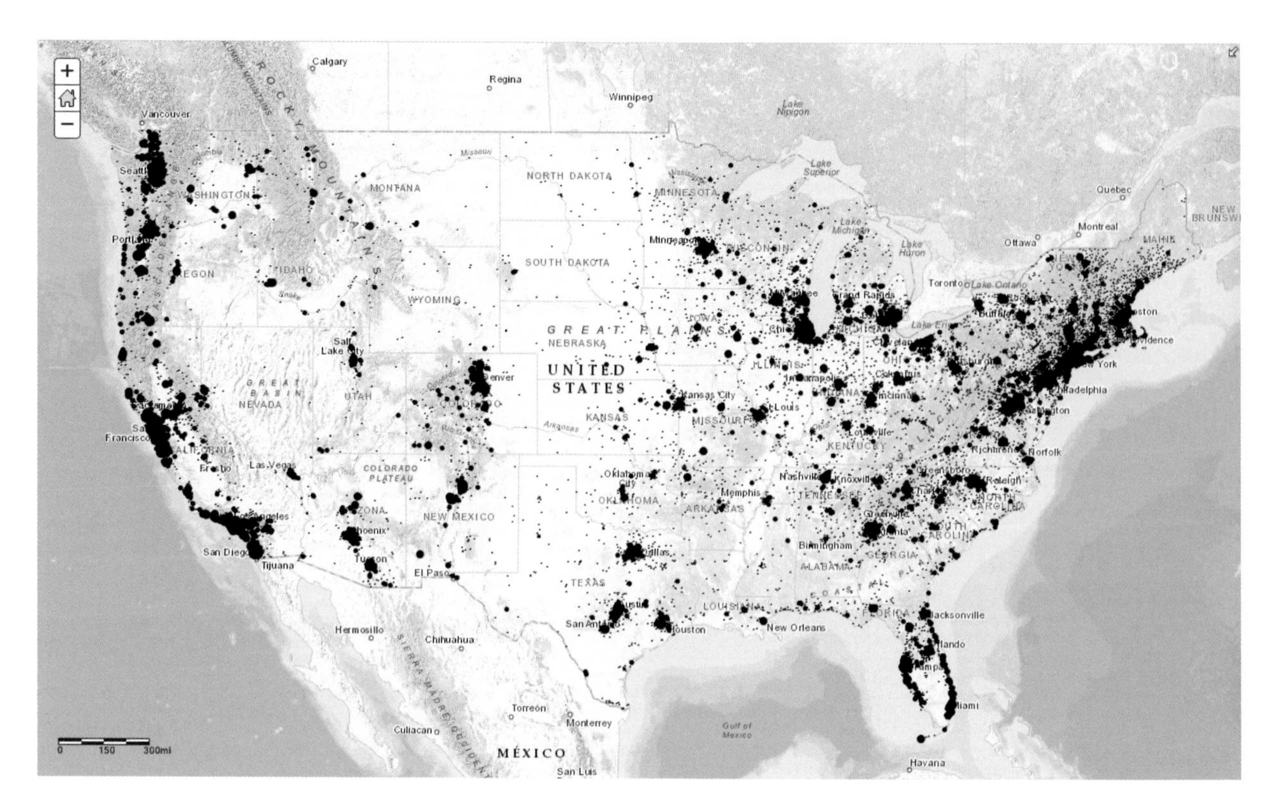

Coding her petition signatures on a GIS map, Madi could finally see where the strongest support for her cause was coming from.

happens to their community resources are the people from those places," she observed. "Oftentimes, they don't have [formal] land rights, and they're using maps that are completely outdated—or they don't know how to use them."

Inspired by the power of community mapping, she created a story map for JGI and Roots & Shoots on Jane's conservation approach, which led to the idea for her undergraduate senior thesis ("Using Geospatial Analysis for High School Environmental Science Education: A Case Study of the Jane Goodall Institute's Community-Centered Conservation Approach"): "to design a mapping case study for high schoolers which illuminates the relationships between social, economic, and ecological needs of a community abroad."

She continues to pursue that interest:

I get to Skype into classrooms frequently and introduce the community mapping model. I've gotten to mentor a lot of kids.

Part of Madi's role as board member of JGI is to increase GIS use with the Roots & Shoots platform, which offers the Tapestry of Hope (http://www.janegoodall.org/?portfolio=putting-hope-map), an online interactive map (powered by Esri® Story Maps® technology). With it, young leaders document the locations and stories of their local projects around the globe. See Roots & Shoots toolkit link: http://www.rootsandshoots.org/toolkit, and Roots & Shoots community mapping link: https://rootsandshoots.org/mapping.

Madi began simply by wanting to save the animals, but to do that, it seemed as if you must save the whole world as well, she says. It's "a lot of responsibility for a kid," Madi says.

These problems are really big, and kids can be overwhelmed. Being able to map your project and see that you're part of a much larger movement helps you gain perspective.

Madi says she feels "just as driven" now as when she was 11 years old. She currently sits on the board of the JGI and serves as committee chair for Roots & Shoots. Roots & Shoots offers a site called Community Mapping in Action with ArcGIS at https://rootsandshoots.maps.arcgis.com/home/index.html, along with tutorials for young people.

I want younger people to have those lightbulb moments, like, Oh, I can apply GIS for x, y, or z—just like I did. That's what excites me.

Madi is always working toward a brighter future—for others and for herself.

In July 2018, Madi received a scholarship to pursue her master's in environmental policy and development at the University of Cambridge and was excited at the prospect of studying where her mentor went to school: "This fall, I'm moving to the UK—I'll be at Darwin College, where Jane went!"

Looking back, Madi says of her Girl Scout campaign:

In my experience, starting the project was exhilarating. It was something I simply had to do. But ending my campaign was scary, because who was I going to be in the world if I wasn't the "Girl Scout cookie girl"?

Madi is grateful for the mentorship of such accomplished women as Maria Pacheco, *right*, a Guatemalan businesswoman and scientist.

Says Madi: "It was the right decision, and just six months later, Girl Scout cookie baker Kellogg and Wilmar, the world's largest supplier, adopted deforestation-free policies, largely because of my campaign's previous momentum."

And yet, Madi doesn't hesitate in giving credit to those who went before:

It's taken the work of generations of women advocating for equality so that I can pursue leadership positions. I'm grateful for trailblazing women, like Jane Goodall, who have pursued careers in science.

The courage and success of these women have helped Madi pursue her own career today, always on the lookout for new challenges. ✷

Acknowledgments

For writing this book, special thanks to the following:

Stacy Krieg Bringing it all together

Stacy Krieg is a senior acquisitions editor at Esri Press. Her experience is as an editor, writer, publisher, and storyteller, and she's passionate about creating content that can help improve people's lives. She feels privileged to get to know the women in this book and honored to help them tell these inspiring stories. A native East Coaster, Stacy now lives in Redlands with her three children, dog, and cat.

Claudia Naber Carving an adventurous spirit

Claudia Naber is an acquisitions editor at Esri Press. She truly enjoyed getting to know the women whose stories she helped craft. Some she had worked with on previous projects, but she had little idea about the life journeys they had taken; others became new inspirations. Claudia is also a caretaker and administrator at Animazonia Wildlife Foundation, a sanctuary for big cats rescued from threatening conditions in captivity or displaced in the wild.

Candace Hogan Acquisitions editor

Also, thank you to the following members of Esri Press for their contributions to this book:

- Carolyn G. Schatz, editor
- Sasha Gallardo, copy editor
- Monica McGregor, designer
- Mike Livingston, market development
- Catherine Ortiz, manager and publisher